Practical Patient Management in Reproductive Medicine

Evidence- and Experience-Based Guidance

Practical Patient Management in Reproductive Medicine

Evidence- and Experience-Based Guidance

David J Cahill MD, FRCPI, FRCOG, FHEA
Emeritus Professor in Reproductive Medicine and Medical Education
University of Bristol, St Michael's Hospital, Bristol, UK

CRC Press
Taylor & Francis Group
Boca Raton London New York

CRC Press is an imprint of the
Taylor & Francis Group, an **informa** business

CRC Press
Taylor & Francis Group
6000 Broken Sound Parkway NW, Suite 300
Boca Raton, FL 33487-2742

Printed on acid-free paper

International Standard Book Number-13: 978-1-138-33562-2 (Hardback)

Visit the Taylor & Francis Web site at
http://www.taylorandfrancis.com

and the CRC Press Web site at
http://www.crcpress.com

Contents

Introduction

This book provides junior doctors and doctors in training and in practice with useful advice on managing women and men who are having trouble conceiving and couples with infertility and related disorders. It is aimed primarily at those taking postgraduate professional examinations in obstetrics and gynaecology and those involved in master's-level studies but should also be useful and accessible to undergraduates, general practitioners and primary care physicians, providing in-depth explanations for conditions in this broad area. How things have changed since 1990 when I first started to work in this field – success rates, the introduction of transvaginal oocyte recovery, the use of GnRH agonists to suppress unwanted ovulation, greater equity for patients in access to treatment and perhaps regrettably the shift of assisted conception from a cottage industry to one with multinational corporate status. I hope that if this book doesn't give you the information you need, it will point to a source you can use.

Throughout the text, I refer to couples – most of the problems in reproductive medicine refer to difficulties between two people, a man and a woman, while of course recognising that there are multiple relationship options for couples to be in. Patients are a great resource: we learn from them, they challenge us with the complexity of their problems and occasionally, the more assertive will directly challenge our approach and management – and usually they are right – do not forget that.

This book is based on a background of extensive research, building on work in all areas of reproductive medicine and surgery by my predecessor Professor Michael Hull and his team and by myself and close colleagues, my 20+ years of experience in running infertility clinics, all in a tertiary university teaching hospital, St Michael's Hospital, Bristol. There, I have gained insight in the concepts that trainees struggle with. The driving force behind this book is to explain and simplify that. Most chapters have a case study or two included to put flesh onto the dry bones of the narrative, to recognise that all situations are not simple or black and white, but that there are subtleties and balances to each clinical decision made. These are composites of patient encounters and relate to no one couple individually.

This book does not attempt to tackle conditions and situations relating to contraceptive and hormone replacement choices, though these are all within the spectrum of reproductive medicine; many books are available to address those areas specifically. I briefly touch on assisted conception to outline the current broad management of patients – and only briefly, as this is well catered to in terms of reading material elsewhere (from techniques on surgical sperm recovery to advanced imaging of embryo development) and, further, the field changes so rapidly. This book is designed to be an accessible handy reference in outpatient clinics and in revision for professional exams as well as a fully referenced information resource. I trust you will find it so.

David J Cahill

Too often all are prone to forget that living people conceive, not ova and spermatozoa (1).

REFERENCE

1. Guttmacher AF. Factors affecting normal expectancy of conception. *Journal of the American Medical Association*. 1956;161(9):855–60.

Acknowledgements

This book is based on the experience of more than 20 years in one infertility clinic, with useful critical input from several individuals: my mentor Peter Wardle, MD, FRCS, FRCOG; my trainee Guy Morris, MRCOG and my children, all of whom read various drafts. I am particularly indebted to the intellectual and editorial input of my wife Eileen, MRCPsych.

I am grateful to all the trainees I have worked with over the years, for what I learnt from them, to my nursing colleagues who have been a joy to work with and to the patients for all the wonderful journeys and learning experiences they have taken me on.

Several people have given invaluable input in Chapter 10: Dr Elsamawal Elhakim, Bahrain; Dr Caroline Fertleman, UCL, London; Dr Uma Kondaveeti Gordon, Bristol.

Glossary of terms

17OH-progesterone: 17-hydroxyprogesterone is produced by the adrenal gland. It is one test undertaken to differentiate between hirsutism and androgenic effects from PCOS as opposed to those from congenital adrenal hyperplasia, which it indicates.

60-kDa heat shock protein: This protein is excreted by the *Chlamydia* bacterium even when it is in a dormant state, and its effect is to induce a low-level inflammatory response in tubal epithelium and cilia, leading to tubal damage.

Add-back therapy: The addition of low-dose continuous hormone replacement in women having long-term oestrogen suppression using a GnRH agonist as part of preoperative planning for fibroid surgery or as part of endometriosis treatment. Its benefit is to minimise the unwanted menopausal symptoms of the GnRH agonist (both symptomatic and asymptomatic).

Add-ons: Additional treatment options, not part of the standard package of IVF and ET, which are offered to couples as part of treatment. Some are evidence based such as assisted hatching; others lack any evidence base and regulatory authorities are concerned about their use, in the absence of evidence, by couples who will agree to anything if it is suggested it will increase success rates.

Adhesions (or synechiae): Strands of scar tissue evident inside and outside the uterus, frequently associated with difficulty in conception.

Adrenal zona reticularis: The innermost layer of the adrenal cortex, closest to the adrenal medulla, where the sex steroids are produced.

Adrenarche: The development of secondary sexual characteristics such as axillary and pubic hair which are determined by adrenal gland activity.

Alkylating agents: A powerful group of chemotherapeutic drugs, damaging to DNA synthesis. Examples from clinical practice include cyclophosphamide, cisplatin and busulfan.

Amenorrrhoea: The complete absence of periods, usually accompanied by anovulation. The most common cause of amenorrhoea is *always* pregnancy.

Amino acid: The fundamental building block of peptides and proteins – an amino acid is at core based on a central carbon atom, with a valency of 4 – and arranged about it will be one basic amino group ($-NH_2$) and an acidic carboxyl group ($-COOH$), with an organic side chain unique to each amino acid. The simplest formula for an amino acid would be CH_2 NH_2 COOH, but, the simplest occurring in nature is CH_3-CH (NH_2)-COOH (alanine).

AMH (anti-Mullerian hormone): This hormone has several actions throughout life: *in utero*, it is secreted to suppress the Mullerian tract from developing in the male. In adult female reproductive life, it has been found to be a marker of ovarian reserve (see later) which is produced by small follicles, before they are receptive to FSH.

Ampulla of the vas deferens: The pre-ejaculatory place in the male reproductive tract, where sperm are held in fructose-rich fluid.

Analysis of DNA fragmentation: See DNA fragmentation.

Androgen binding protein (ABP): A protein found in the lumen of the seminiferous tubule, whose function is to bind and concentrate androgen activity.

Anejaculatory: The inability to ejaculate at orgasm for men, generally either due to neurological or anxiety states.

Anovulation: The failure of ovulation to occur, for whatever reason (See Sections 5.1 and 5A.1 in Chapter 5).

Anteroventral periventricular nucleus: An area of the hypothalamus responsible for the release of neurotransmitters that control GnRH activity in puberty initially and then through reproductive life.

Antioxidant therapy: A putative therapy to overcome oxidative stress in men with sperm disorders, of uncertain value as yet.

Arcuate nucleus: An area of the hypothalamus responsible for the release of GnRH activity in puberty initially and then through reproductive life.

Aromatase inhibitors: Aromatase is the enzyme in the ovary which converts androgens to oestrogen. Inhibitors are used to suppress oestrogen production temporarily to induce the HPO axis to secrete higher levels of FSH to stimulate ovarian follicles.

Asherman's syndrome: A condition recognised by amenorrhoea which tends to result from damage to the endometrium, usually following surgery for miscarriage.

Assisted conception: The techniques involved in the artificial process of initiating a new pregnancy, with success rates usually considered as 'take-home baby rates'.

Assisted hatching: A technique to improve implantation rates, often carried out using laser, and aiming to remove a portion of the zona pellucida.

Avon Longitudinal Study of Parents and Children (ALSPAC): Also known as Children of the 90s, Prof Jean Golding initiated this longitudinal study of all children born in Bristol (Avon), England, over a 2-year period in 1991–1992, which has afforded access to incredible amounts of data across all aspects of medicine. The children are now in their mid-20s and many are still participants.

AZFc: One of three azoospermia zone fractions of the Y chromosome, all of which lead to azoospermia or severe oligozoospermia. AZFc is the abnormality is the only one where there is a (very small) possibility of sperm production which can be collected using surgical sperm recovery.

Azoospermia: The complete absence of sperm in the ejaculate.

Blastocyst transfer: Transfer of a 32–64 cell structure following fertilisation and zygote development. It differs from "embryo transfer" which is usually related to a 4–6 cell structure. If the embryo grows on to 32–64 cells, it is generally more likely to implant and give rise to a pregnancy

BMI: See Body mass index.

Body mass index (kg/m²): A method of describing an individual's weight related to their height. It is calculated by taking weight in kilograms and height in metres, and then the BMI is in kilograms per metre (in height) squared – kg/m^2.

Bone age measurement: A radiological test undertaken on the wrist bones, as these tend to calcify in a set and predictable pattern by age, allowing accurate estimate of bone age from an x-ray of the wrist bones.

Cannabinoids: Chemicals released by cannabis flowers which bring about the effects of cannabis.

Central precocious puberty: A condition due to the early onset of activity in the arcuate nucleus and thus early onset of gonadotrophin secretion. It can be idiopathic or secondary to other factors (see Section 1.2.1 in Chapter 1 for details).

Cervical mucus: The fluid which is produced by the glands in the cervix, which itself has varying biological properties during the menstrual cycle, thick and impervious to sperm during most of the follicular and luteal phases, changing to thin, runny and permissive to sperm passage at midcycle in response to the oestrogen surge that triggers the LH surge.

Cervix: The lower part of the uterus, consisting mostly of connective tissue, with a canal through it from the vagina to the uterine cavity, in which are found the most architecturally interesting crypts along which sperm can swim to facilitate their passage up to the uterine cavity. The opening to the vagina is the site of the squamo-columnar junction (the epithelial lining of the vagina and the cervix), which is the site of most cervical cell abnormalities leading to carcinoma *in situ* in the cervix.

Chemotherapeutic drugs: A wide range of drugs used to kill actively dividing cells – usually cancer cells.

Chlamydia: *Chlamydia trachomatis* is a sexually transmitted disease. It is an obligate intracellular organism (it cannot survive outside a cell) and it causes fimbrial and cilial damage to a fallopian tube when infected. It is likely that more than one episode of infection is required to bring about this damage (see Section 6.4 in Chapter 6 for details).

Chlamydia antibody blood test: A blood test which can measure the amount of antibody in the blood to Chlamydia antigen, found on the surface of the Chlamydia organism. It can accurately predict the probability of a woman having any or severe tubal damage (see Figure 6.3).

Chromosomes: The 46 tiny strands of genes that are contained in every cell in the body and control their function. Twenty-three chromosomes are inherited from each parent. Down's syndrome is an example of a chromosomal abnormality – in this condition, there are three copies of chromosome 21, as opposed to the usual number of 2.

Clomifene citrate: One of the simplest and least effective of the fertility drug family (see Section 5.3 in Chapter 5 for details).

Computer-assisted sperm analyses (CASA): A complex form of sperm analysis, in which sperm movement is assessed for features such as lateral head displacement and curvilinear velocity. Its advocates consider CASA superior to other sperm analysis techniques.

Conception: The process of oocyte and sperm fusing to become a zygote, and with continued growth thereafter. Often used interchangeably with 'getting pregnant'.

Congenital adrenal hyperplasia: A cause of precocious puberty, in which excessive androgens are produced. This is usually caused by a deficiency of the enzyme 21-hydroxylase. Treatment with glucocorticoids will protect from adrenal insufficiency and suppress the excessive adrenal androgen production.

Constitutional delay: Most (~50%) children with delayed puberty have constitutional delay, whereby there is no specific cause or treatment.

Contrast ultrasonography: A technique using ultrasonographically translucent fluid inserted into the uterine cavity to outline it and the fallopian tubes without the use of ionising radiation and with the advantage of being used without concern throughout the menstrual cycle. No RCT exists to show which is better, this technique or a hysterosalpingogram, and the choice remains individualised.

Cryopreservation: The technique of freezing biological material whereby cryoprotectants are added to the material to prevent cytoskeletal damage (see later) and then the material is frozen to very low temperature levels (usually less than −200°C).

Cystic fibrosis: A condition which is genetically inherited through autosomal recessive inheritance. The disease occurs when both copies of a gene (ΔF508) for a specific protein (cystic fibrosis transmembrane conductance regulator) are malfunctions, leading to increased thickness of mucus secretions. This causes mucus buildup in all the affected organs – the lungs, the pancreas, the bowel. Most males with the condition will have congenital bilateral absence of the vas deferens and will be infertile as a consequence. Most common in Northern Europeans and particularly those of Celtic extraction, there is no cure and treatments are supportive. Life expectancy for children born with this condition in the developed world is now well into their 40s.

Cytoskeleton: The skeletal structure in cells which at the time of fertilisation and mitosis pulls chromosomes apart to assist in the process of cell division. Of relevance when freezing cells for future use, the cytoskeleton can be easily disrupted by cell freezing or thawing.

Delayed puberty: When puberty has not occurred more than two standard deviations later than the average age, which is beyond 13 in girls, 14 in boys.

DNA fragmentation: A sperm dysfunction in which the DNA becomes broken up and disintegrates, leading to the potential for infertility, as the DNA in the chromosomes does not function correctly and this leads to fertilisation difficulty and failure of IVF to lead to a pregnancy. Use of tests for DNA fragmentation in clinical practice is debated with regard to their usefulness.

Donated gametes: Sperm or oocytes from another person to use in trying to get pregnant; oocytes used are often fresh, sperm generally used frozen.

Donor insemination (DI): The use of sperm from another person, generally anonymous, screened for infections and provided to women in a clinical setting inserted into the uterine cavity, often used from freeze storage.

Downregulation: A chemical process, which refers to the diminution of receptors to GnRH on the surface of gonadotroph cells in the pituitary. It leads to the reduction in numbers of receptors so that any endogenous GnRH has much less opportunity to find and stimulate a receptor, thus amplifying the effect.

Dynorphin A: A neurotransmitter kinin peptide hormone which is an important regulator of GnRH hormone and a suppressor of its secretion at the onset of puberty until Neurokinin B overcomes it.

Ectopic pregnancy: A pregnancy which occurs in any site outside the uterine cavity, the most common in the fallopian tube in the isthmus or ampulla (~75%) (see Figure 6.8 for an illustration of where the isthmus and ampulla are).

Egg (oocyte): See Ooycte.

Ejaculatory dysfunction: Any form of disorder of the ejaculatory process, from premature to anejaculation. This is caused by a wide variety of reasons – anxiety, ingested drugs or disease processes such as diabetes.

Embolisation: Any process of blocking a blood vessel, but in reproductive medicine, specifically this refers to the catheterisation of the uterine artery and the blockage of vessels feeding a fibroid to bring about necrosis and some subsequent shrinkage (rarely more than 50% of the original size).

Embryo: Widely used loosely to describe the morula/blastocyst just after fertilisation – the embryo stage doesn't actually occur until 2 weeks later.

Embryo freezing: The process of cooling and removing fluid from a morula/blastocyst in so that it is brought to a very low temperature and kept at that temperature until it is time to thaw and replace it (see Section 8.4.4 in Chapter 8).

Endometrial atrophy: The effects on the lining of the uterus after menopause, in which the endometrium thins and appears pale and flat at hysteroscopy.

Endometriosis: A condition whereby tissue resembling endometrium, with glands and stroma, is found outside the uterine cavity. Within the uterine body, this is called adenomyosis; outside, it is called endometriosis. It would appear to be responsive to hormonal stimulation and is found as a pathology most commonly in the pelvic cul-de-sac (pouch of Douglas) and behind the ovary. Distal sites are rare but fascinating – e.g. in the lungs, thighs – and can still cause symptoms.

Endometrium: The interior lining of the uterus, responsive to hormone stimulation every menstrual cycle, and shed at every menstruation. It has a key role in implantation.

Epididymis: Site of storage and maturation of sperm in their development (see Figure 4.3), though they are still incapable of forward motility on entry to the epididymis. This is the site of main storage of sperm until ejaculation gaining forward motility during their time in the epididymis.

Epiphysis/eal: This structure is a part of long bones, lying adjacent to the metaphysis with the epiphyseal plate in between. This is the site of all bony growth in teenage years, and if closed too early by inappropriately high levels of hormones, further bone growth cannot occur.

Fallopian tube: The structure connecting the uterine cavity and the ovary, which runs from the fimbriae to the cornual end, the central lumen lined with cilia, and divided traditionally into four sections (see Figure 6.8).

Fertilisation: The fusion of a male and female gamete (sperm and oocyte) to form a new biological organism, the zygote. This zygote then moves almost instantly into a mitotic division, leading to further multiplication until the 16-cell stage (3–4 days later), the blastocyst and then at 2 weeks later, the embryo.

Fibroid: See Myoma.

Fibromyoma: See Myoma.

Fimbria(e): The end of the fallopian tube furthest from the uterus and closest to the ovary. The fimbriae reach out to the ovary under the direction of chemotaxis, possibly mediated by platelet-derived growth factor (see https://doi.org/10.1371/journal.pone.0158266). Damage to fimbriae by infections such as Chlamydia and gonorrhoea can lead to fimbriae adhering to each other, leading to loss of function and, in severe cases, to closure of the fimbrial end of the tube, leading to a hydrosalpinx (see Hydrosalpinx).

Follicle: In the ovary, the follicle is the centre of endocrine activity in oocyte development and maturation. Follicles develop receptors to FSH when they are approximately 5 mm in diameter, and they then grow and develop to about 20 mm in size when they are producing maximal oestradiol amounts and are ready for ovulation.

Follicle count (AFC): A measure of ovarian reserve, by assessment of the numbers of follicles on ultrasound between 2 and 6 mm in the first week after menstruation.

FSH: A glycoprotein secreted in the pituitary gland, with an α and a β chain. The α chain is common to the other pituitary glycoproteins (FSH, LH, ACTH) and to hCG; the β chain gives FSH its specificity in action and structure, which are: stimulation of the granulosa cells to produce oestradiol, increasing the follicular fluid volume to ensure the follicle increases.

Full blood count and ferritin levels: Investigations done as part of the routine workup for couples trying to conceive – low haemoglobin levels and low iron stores, indicated by ferritin levels, are a suboptimal way to begin a pregnancy, and it makes sense to ensure that these are optimal beforehand, There is some poor-quality evidence that low ferritin levels are associated with low fertility (https://doi.org/10.1016/0140-6736(91)93255-8).

Gamete: A haploid cell (with half the usual number of chromosomes), ready for fertilisation, containing all the chromosomal information to meet with its counterpart (sperm or oocyte) to form a new embryo with its own unique genetic components.

Gamma-aminobutyric acid (GABA): A neurotransmitter which, in the anteroventral paraventricular nucleus, acts to retard the onset of puberty, until kisspeptin stimulus overcomes it.

Glutamate: The anion of glutamic acid in its role as a neurotransmitter, of importance in reproductive biology, as it has a stimulatory influence in the initiation of puberty.

GnRH: See Gonadotrophin-releasing hormone.

GnRHa (gonadotrophin-releasing hormone agonists): Synthetic analogues of GnRH, with substitutions in the amino acids outlined in **bold** to give therapeutic analogues, Glu-His-Trp-Ser-Tyr-**Gly**-Leu-Arg-Pro-**Gly**. The action of these agonists is to cause flare-up and stimulation, downregulation of receptors, desensitisation of gonadotrophin cells, and then suppression, and they are slowly reversible.

GnRHant (gonadotrophin-releasing hormone antagonists): Synthetic analogues of GnRH, with substitutions in the amino acids outlined in **bold** to give therapeutic analogues, **Glu-His-Trp**-Ser-Tyr-**Gly**-Leu-**Arg**-Pro-**Gly**. The action of these agonists is to cause receptor blockage (probably without downregulation); they are competitive inhibitors at the receptor level, cause immediate suppression and are rapidly reversible. They have only recently come into therapeutic value, as previous compounds had unwanted histamine side effects.

Gonadal dysgenesis: Any failure of reproductive organ (gonad) in intrauterine life – Turner's syndrome is the most well known of these – streak gonads are a general finding in gonadal dysgenesis.

Gonadarche: The initiation of function of the gonads at puberty (see Section 1.1 in Chapter 1 for details).

Gonadotrophin: A stimulatory hormone from the anterior pituitary (endogenous) or injected (exogenous) used to stimulate the ovaries and testes to produce oestrogen and testosterone.

Gonadotrophin-releasing hormone: The hormone arising from the arcuate nucleus responsible for stimulation of FSH and LH from the anterior pituitary, a decapeptide whose amino acid sequence is pyroGlu-His-Trp-Ser-Tyr-Gly-Leu-Arg-Pro-Gly-NH_2.

Gonadotrophin-releasing hormone analogue family (see also GnRHa): A family of drugs used extensively in gynaecology, with the general intent of causing suppression of oestrogen production. They are consequently valuable in ovulation induction and superovulation for IVF, and ovarian suppression as preoperative treatment for fibroid/myoma removal. They are synthesised by replacement of one or two amino acids in the ten amino acid (decapeptide) natural hormone to bring about significant lengthening of the duration of action, and pharmacologically the effect is to suppress and reduce the number of receptors for GnRH so that its effect is reduced dramatically, and the receptor's numbers are downregulated.

Gonorrhoea: A sexually transmitted infectious disease, caused by *Neisseria gonorrhoea* (a.k.a. the clap), a gram-negative intracellular diplococcus which presents as urethritis and urethral or vaginal discharge, and can be asymptomatic. It is passed by body fluid contact, and any sexual contact can lead to this – though the schoolboy myth of toilet seat contact is just a myth.

Grays: The unit of absorbed radiation, equivalent to 1 Joule/kilogram.

Growth spurt: The growth in height that occurs under the stimulus of oestrogen/testosterone at the time of puberty which gets arrested by close of the epiphyses (at the end of long bones).

hCG (human chorionic gonadotrophin): A glycoprotein secreted in the syncytiotrophoblast, with an α and a β chain. The α chain is common to the other pituitary glycoproteins (FSH, LH) and to hCG; the β chain gives hCG its specificity in action and structure: initial rescue of the corpus luteum to maintain progesterone production until the placenta takes over, several weeks later.

Hepatitis B and C antibodies: Hepatitis B and C are both viral infections affecting the liver, which are transmitted through body fluids and therefore by intravenous drug injections, by sexual intercourse and transplacentally (C only) or at delivery (in the vagina) (B & C), and are treatable by antiviral therapies.

HIV 1 and 2: Generally both tested for as part of investigations for infertility, HIV-1 and HIV-2 are the two main types of HIV. HIV-1 is widespread worldwide. HIV-2 is found in Western Africa, and is a less pathogenic type.

hMG (human menopausal gonadotrophin): A stimulant of follicular cells derived from the urine of postmenopausal women; subject as a result to ebbs and flows in supply and to concerns occasionally about contamination (in the past, with prions).

Hormone: Any chemical which is released at one site and has its action elsewhere (endocrine), at the same site (apocrine) or a short distance away (paracrine).

HSG (hysterosalpingography): A radiological technique using radio-opaque dye to define the cavity of the uterus and the fallopian tubes, out to the fimbrial end.

Human Fertilization and Embryology Authority: The independent regulator in the UK of all fertility treatment using gametes (including IUI) and of research using human embryos.

Hydrosalpinx (pl. hydrosalpinges): A distortion or swelling at the fimbrial end of the fallopian tube – resulting in blockage of the tube and its filling with fluid, which is known to adversely affect the ability of embryos to implant. See Figure 6.4 for a radiological appearance and Figure 6.6 for a diagrammatic representation of a hydrosalpinx.

Hypergonadotrophic hypogonadism: Failure of the gonads to function in the presence of more than enough gonadotrophin stimulation, generally because the gonads have run out of responsive material – the finding when menopause happens and all the oocytes and follicles (with receptors to FSH) are exhausted.

Hypogonadotrophic hypogonadism: The gonads are unstimulated as a consequence of failure of gonadotrophin production. Extremely low levels of FSH and LH should prompt you to consider this diagnosis.

Hypopituitarism: A failure of normal function, by underperforming, of the pituitary gland, often due to a nonsecreting tumour which presses on the rest of the gland.

Hypothalamic-pituitary-adrenal (HPA) axis: The particular relationship between the paraventricular nucleus in the hypothalamus, the anterior pituitary gland, and the adrenals. Responsible for cortisol secretion, it controls mood and emotions, sexuality, the digestive and immune systems and energy storage and expenditure.

Hypothalamic-pituitary-gonadal (HPG) axis (ovary or testis): The particular relationship between the arcuate nucleus in the hypothalamus, the anterior pituitary gland and the gonads (either ovary or testis) with outgoing secretion of GnRH, then FSH and LH and oestrogen/progesterone or testosterone, and those feeding back to the pituitary and to the hypothalamus.

Hypothalamus: A midline structure of the forebrain, below the thalamus, responsible for sleep and emotional control, pituitary stimulus and control and homeostasis (of hunger and temperature, for example).

Hypothalamic hamartoma: A rare cause of precocious puberty, linked with gelastic, focal or tonic–clonic seizures.

Hypothyroidism: Underfunctioning of the thyroid gland, diagnosed by low levels of TSH and a cause of amenorrhoea and oligomenorrhoea.

Hysterectomy: The surgical removal of the uterus and cervix, not often done in reproductive medicine.

Hysterosalpingogram: A radiological technique using radio-opaque dye to define the cavity of the uterus and the fallopian tubes, out to the fimbrial end (see Figure 6.4).

Hysteroscopy: A technique to endoscopically visualise the cervical canal and the uterine cavity, used in reproductive medicine to confirm the normality of the cavity of the uterus.

Idiopathic hypogonadotrophic hypogonadism: The gonads are being unstimulated as a consequence of failure of gonadotrophin production, for no obvious cause. Extremely low levels of FSH and LH should prompt you to consider this diagnosis.

Immature oocyte cryopreservation: One of several techniques to preserve fertility in women who want to preserve fertility prior to chemotherapy. This involves either aspirating small follicles which are unstimulated or taking a piece of ovary and dissecting the oocytes from the ovarian tissue. Presently, it is not considered a very efficient method of preserving oocytes (doi:10.1007/s10815-013-0112-0).

Infertility: The inability to conceive, usually considered to be in the timeframe of 2 years of actively trying to conceive.

Inhibin A: Inhibin was discovered in the 1970s and is recognised to downregulate and suppress FSH production in girls. Inhibin A is most active in the midluteal phase.

Inhibin B: Inhibin is recognised to downregulate and suppress FSH production in girls. Inhibin B is most active in the midfollicular phase. In boys before puberty, this is formed in the Sertoli cells in the testis. It is produced after puberty in response to sperm numbers and FSH, but the mechanisms are not clear – inhibin B is inversely related to FSH and directly related to sperm numbers.

Insemination: The process of getting sperm directly into the uterine cavity/in proximity to oocytes after oocyte recovery/the effect of natural intercourse.

Interstitial cell stimulating hormone: An alternative older name for LH, this hormone in men stimulates the interstitial cells (Leydig cells) to produce testosterone.

Interstitial/corneal end: The end of the fallopian tube closest to the uterus (corneal), and indeed that portion that lies within the myometrium itself (interstitial), connecting to the uterine cavity.

Intracytoplasmic sperm injection (ICSI): A refinement of *in vitro* fertilisation, whereby oocyte insemination is undertaken by direct injection of the oocyte with a single sperm, of particular value in men with very low sperm counts (see Figure 8.2).

Intrauterine adhesions: Scar tissue within the uterine cavity, usually resulting from trauma (overactive curettage after a miscarriage) or from low-grade infection (particularly tuberculosis).

***In vitro* fertilisation treatment (IVF):** The mainstay of assisted reproduction therapies, barring natural cycle IVF (see below):

- Stimulation using GnRH-a + FSH/HMG + USS monitoring
- Oocyte retrieval (see Figure 8.1)
- Inseminating the oocytes in the laboratory
- Culture for 2–3 days – or more
- Embryo (4–6 cells or more advanced) transfer

Intrauterine insemination (IUI): The simplest technique of assisted reproduction, in which prepared sperm are introduced into the uterine cavity. This can be in unstimulated cycles (recommended by NICE but in practice very ineffective) or with mild stimulation with clomifene and/or gonadotrophins.

Kallmann syndrome: Anosmia and hypogonadotrophic hypogonadism, caused by failure of migration of the neurones that will become the olfactory nerve and the arcuate nucleus from the posterior cerebrum forward to their correct position in adult life. The absence of smell is a key to the diagnosis.

Kartagener's syndrome: Also called primary ciliary dyskinesia, this condition is characterised by cilial dysfunction affected the respiratory tract, the fallopian tube and perhaps sperm. Affected individuals have frequent upper and lower respiratory tract infections, often affecting life expectancy. If adult life is achieved, women will tend to be infertile as a result of collection of mucoid material blocking the fallopian tubes, and IVF can circumvent this, if the woman can undergo the surgical procedure to collect the oocytes and is considered physically well enough to be able to undergo pregnancy.

Karyotyping: A laboratory investigation to elucidate the chromosomal content of a cell or cells.

Khat: Khat or qat is a plant native to the Horn of Africa and the Arabian Peninsula. It is chewed to achieve excitement, loss of appetite and euphoria and has unwanted effects on reproductive function (see Section 10.5 in Chapter 10).

Kisspeptin: An important initiator of puberty, kisspeptin, made in the hypothalamus, may also help stop the spread of cancer. It was identified in 1966 in a laboratory in Hershey, Pennsylvania, the home of Hershey's Kisses (and the laboratory was sponsored by Hershey's Chocolates, apparently).

Laparoscopic oocyte collection: Approach to oocyte collection, using laparoscopy to visualise the ovaries and directly aspirate the follicles. Went out of favour when the transvaginal approach was introduced, as the latter requires light anaesthesia and is associated with much less surgical risk.

Laparoscopy (and dye hydrotubation): Laparoscopy is an operative procedure, used first in humans in 1910 and promoted for use in gynaecology by Patrick Steptoe (of Edwards, Purdy and Steptoe and Louise Brown IVF fame) in 1967. It involves insertion of a 10-mm or larger trochar into the umbilicus, with prior insertion of CO_2 to distend the abdominal cavity. A frequently performed test in gynaecology is to inspect the fallopian tubes and, through the cervix, insufflate the uterus and the fallopian tubes with methylene blue dye to inspect the tubes for patency.

Leydig (previously interstitial) cell: Cells which are intimately related to the seminiferous tubules, though immunologically separate from them, which are stimulated by luteinising hormone to produce testosterone, which then migrates across to the Sertoli cells and induces receptor formation in them so that stimulation by FSH induces sperm cell production (see Figure 4.2).

Luteinising hormone (LH): A glycoprotein secreted in the pituitary gland, with an α and a β chain. The α chain is common to the other pituitary glycoproteins (FSH, LH, ACTH) and to hCG; the β chain gives LH its specificity in action and structure, which are: reawakening of the meiotic division process in the oocyte to bring it to a haploid chromosomal state ready for fertilisation, to change the function of the follicle from producing largely oestrogen to producing largely progesterone, to cause the follicle to rupture and release the oocyte and finally to bring about a distinct yellowish-red colour in what was the follicle, now the corpus luteum (yellow body) (lutein = yellow in Greek).

Mayer–Rokitansky–Küster–Hauser syndrome: A rare condition of failure of Mullerian development, resulting in failure of some or all of the Mullerian origin genital tract to develop (upper third of the vagina, cervix, uterus), seemingly of genetic origin, autosomal dominant and linked to abnormalities in chromosomes 1, 4, 8, 10, 15, 16 and 22.

McCune-Albright syndrome: Polyostotic fibrous dysplasia, *cafe-au-lait* skin pigmentation, and peripheral precocious puberty.

Menarche: The onset of menstruation, in puberty. In most girls in the developed world, this should happen around 12.5 to 13 years.

Menopause: The final cessation of menstruation, signifying the loss of ability to conceive, with frequently accompanying symptoms such as hot flushes, mood swings and symptoms related to oestrogen loss – initially vaginal dryness being the most evident.

Miscarriage: Failure of a pregnancy to progress beyond 24 weeks. It can present as bleeding and pain (most commonly), usually preceded by amenorrhoea, or as just amenorrhoea, and the pregnancy dies and is resorbed slowly without bleeding; this rarely occurs in pregnancy beyond 6 weeks gestation.

Mitochondrial abnormalities: Mitochondria are intracellular structures that produce the energy (ATP) with the cell. When they malfunction, energy production will fail, but 90% of mitochondrial function is unrelated to energy production and so failure can lead to other problems – haemoglobin, neurotransmitter and cholesterol metabolism disorders as examples. Developmental delay and learning disabilities are some of the ways in which mitochondrial diseases can present.

Mullerian agenesis: The complete failure of the Mullerian ducts to form in a female. See Mayer–Rokitansky–Küster–Hauser syndrome.

Mullerian anomaly: Any abnormality of the Mullerian duct development, which can be as severe as atresia but also leading to duplex upper vagina, cervix and uterus – with uterine anomalies being part of the spectrum.

Mullerian tract: In the early (less than 12 weeks) development of the embryo, the mesonephric ducts (destined to be the male Wolffian ducts) and the paramesonephric ducts (destined to be the Mullerian ducts) develop at the anatomic level of the early kidneys, Mullerian ducts being lateral to the Wolffian ducts. In males, the testes produce AMH to suppress Mullerian tract development. In females, the distal Mullerian tracts unite in later embryonic life to form the uterus, cervix and upper third of the vagina.

Multiple pregnancy: The state of having more than one gestational sac and pregnancy (usually in the uterus), though I once encountered a woman who had two sacs, one on the right and one on the left fallopian tube, both removed within a week of each other (https://www.ncbi.nlm.nih.gov/pubmed/19835514). Multiple pregnancy is associated with higher risk of both early pregnancy and late pregnancy problems and has been the subject of a UK campaign to reduce multiple pregnancy to the minimum – most clinics now only transfer one embryo at a time in women under 35 years.

Mycoplasma: An uncommon sexually transmitted disease, *Mycoplasma genitalium* causes male and female urethritis, female cervicitis and pelvic inflammatory disease.

Myoma: A benign tumour of the muscle of the uterus, also called a fibromyoma or leiomyoma or fibroid (in my career, I have only seen one woman whose myoma was malignant) – growing generally under oestrogen stimulus to be anything from pea-sized to melon-sized or bigger. Symptoms are sometimes caused by size and by uterine cavity distortion causing heavy periods.

Myomectomy: Surgical (open or laparoscopic) removal of one or more fibroids, anecdotally associated with extensive haemorrhage and adhesions and risk of hysterectomy.

Myometrium: The muscle layers of the uterus, which are thicker in the upper segment, thinner in the lower segment (that part which lies between the reflection of the bladder at surgery).

Natural cycle IVF: The use of *in vitro* fertilisation without stimulatory drugs to increase oocyte numbers and without timing of LH release by exogenous hormone, relying totally on the woman's own FSH to stimulate follicle growth and LH to trigger ovulation. Likely to give rise to lower pregnancy rates, it is promoted as a less stressful form of treatment, with reduced risk of multiple pregnancy, and may as a consequence be quite suited to some people's needs or wishes from treatment. My own experience is of a high-intensity, low-success rate treatment.

Neurokinin B: A neurotransmitter kinin peptide hormone which is an important regulator of GnRH hormone and initiator of its secretion at the onset of puberty (kinins generally induce vasodilation and smooth muscle contraction).

Obese: The condition of being heavier in weight than desired defined by having a body mass index (in kilograms per metre [in height] squared – kg/m^2) over 30 kg/m^2. In this category, the likelihood of getting pregnant falls significantly (see Section 2.1.2 in Chapter 2).

Oestradiol: One of the oestrogens – see Oestrogen.

Oestrogen: The generic term for a sex hormone (of which there are several) with the properties of developing and maintaining the female characteristics of the body (body shape, skin quality, breast and endometrial function, bone mass). These are all 18-carbon atom structures derived from cholesterol – the three key oestrogens in reproductive life are oestradiol (in normal day-to-day life from 12 to 50), oestriol (produced by the placenta and important in pregnancy) and oestrone (produced in postmenopausal life, after the age of 50 usually). They differ only slightly by having two OH groups, three OH groups and a ketone (respectively) attached to the ring structure of the hormone.

Onan: A biblical character who married his dead brother's wife as was his duty but didn't want children with her (therefore, he withdrew before ejaculation, 'spilling his seed'). Because this was seen as a dereliction of his duty, he was struck dead.

Oocyte: The female analogue of a sperm: this is a haploid cell (with 23 chromosomes, including one X [sex] chromosome) which is ready for fertilisation by a sperm and is released from the ovary by the action of luteinising hormone into and down the fallopian tube, to meet (or not) a sperm ready to fertilise it.

Ovarian hyperstimulation: A process in which gonadotrophin injections are given to stimulate several follicles in the ovary, with the stated intention of producing several eggs for *in vitro* fertilisation (as opposed to ovulation induction).

Ovarian hyperstimulation syndrome: Consisting of stomach bloating and abdominal pain, with some nausea, when severe consisting of haemoconcentration, oliguria or anuria, ascites, risk of DVT or pulmonary emboli, and this latter is life-threatening. Now much rarer than 20 years ago, as treatment plans have been altered to recognise its potential and move to different medications, especially for luteal support (see Section 8.7.1 in Chapter 8).

Ovarian reserve: Tests are available to see what potential remains in the ovary for ovulation, oocyte release and reproduction. These tests includes blood tests such as inhibin, FSH and anti-Mullerian hormone, and ultrasound tests such as the antral follicle count (see also Section 3.5.1 in Chapter 3).

Ovarian tissue cryopreservation: The process of storing some (or less commonly all) ovarian tissue to maintain or preserve some potential for reproduction in cases of cancers likely to destroy ovarian tissue.

Ovary: The female organ situated (normally) in the sides of the pelvis, attached to the uterus by the ovarian ligament. The ovary contains all the germ cells of a woman, which are of the order of 400,000 at the onset of reproductive life at puberty. The ovary is responsible for the production of almost all the oestrogens produced during active reproductive life.

Overweight: The condition of being heavier in weight than desired defined by having a body mass index (in kilograms per metre [in height] squared – kg /m^2) between 25 and 29.9 kg/m^2. In this category, the likelihood of getting pregnant falls, though not significantly.

Ovulation: The process of release of an mature oocyte primed by oestradiol and stimulated by LH to reawaken meiotic division.

Ovulation failure: At its simplest, ovulation is failing to occur, but generally it has a more serious/sinister meaning of ovulation coming to a halt because the number of oocytes available to be recruited has fallen to a level that it no longer occurs and reproductive life is over.

Ovulation induction: A process in which gonadotrophin injections are given to stimulate one (or two at most) follicles in the ovary, with the stated intention of producing one egg for insemination by intercourse or intrauterine insemination (as opposed to superovulation).

Ovulatory dysfunction: Any disorder of the ovulatory process which leads to irregular or absent ovulation – the most common are PCOS, hyperprolactinaemia and hypothyroidism (see Chapter 5, Appendix).

PCO (polycystic ovaries): A contentious and ill-defined condition established using ultrasonography; it generally requires a number of features, a peripheral ring of immature follicles in the ovary (6, 12 or more) with increased stromal thickness and an ovary which is larger than usual (see also Polycystic ovarian syndrome [PCOS]).

PCT (postcoital test): A validated test of sperm and coital function, which is falling away from use. It requires little technological input and perhaps that has led to its demise, as it appears subjective and quality control seems poor.

Peptides: Peptides are short chains of between 2 and 50 amino acids (see Amino acids).

Peripheral precocious puberty: See also Pseudoprecocious puberty and Chapter 1.

PGD: See Preimplantation diagnosis (PGD).

PGS: Similar in technique initially to PGD, preimplantation genetic screening (PGS) involves screening of the entire genetic material in chromosomes. Aneuploid (incorrect number of chromosomes) embryos generally result in a miscarriage or may lead to the birth of a child with a genetic condition. Only embryos which have a normal chromosome content are transferred to increase the chance of success.

Pituitary: A pea-sized organ situated in the sella turcica, found by considering the intersection of two lines – one directly back from the bridge of the nose, the other directly across the head from the front of the ear, just above the external auditory meatus. It is a critical organ in endocrine control in the human – the anterior pituitary (controlled by blood-borne controlling hormones from the hypothalamus) secretes six different hormones, though several are glycoproteins, close in structure with common alpha chains (TSH, FSH and LH); the three others are adrenocorticotrophic hormone (ACTH), prolactin and growth hormone (GH). The posterior pituitary, controlled by neurotransmitters reaching down into the posterior pituitary from the paraventricular nucleus in the hypothalamus, secretes two hormones, vasopressin and oxytocin – controlling blood pressure and fluid retention in the kidney (vasopressin) and milk release on suckling (oxytocin). Oxytocin has some properties of vasopressin, particularly when given exogenously intravenously.

Placenta: The organ that develops from syncytiotrophoblast and cytotrophoblast by 10–12 weeks gestation to be the means of transfer of oxygen and nutrients into the fetus and of carbon dioxide and waste products to the maternal blood. It is highly specialised and affected by many maternal states including hypertension, smoking and drugs ingested – some drugs can pass across the placenta, others not, depending on size – most with a molecular weight less than 500 Daltons will cross.

Polycystic ovarian syndrome (PCOS): The revised 2003 criteria (*Fertil. Steril.* 2004: 81(1) 19–25) suggest that at least two of the following three criteria need to be met to warrant the diagnosis of polycystic ovary:

1. Oligo- or anovulation
2. Clinical and/or biochemical signs of hyperandrogenism
3. Polycystic ovaries and exclusion of other aetiologies (congenital adrenal hyperplasia, androgen-secreting tumours, Cushing's syndrome)

Polyvinyl alcohol (PVA) particles: The material used in transarterial fibroid embolisation.

Precocious puberty: Puberty occurring at a very early age (more than two standard deviations from the norm, and usually considered before 8 for a girl and 9 for a boy). In most cases, unlike pseudopuberty, the process is normal in every aspect except the unusually early age, representing a variation of normal development. Occurring too soon can have unwanted effects on ultimate height gain, and may need to be suppressed as a consequence.

Preimplantation diagnosis (PGD): A test/investigation undertaken, usually on blastocysts, whereby one cell is removed and analysed for its genetic components, usually done to look for a specific gene or abnormality, and, if found, generally that blastocyst will not be transferred back to the endometrial cavity, to avoid a child suffering from that abnormality. It raises ethical questions about whether it is right in principle to screen like this and in particular, as the fate of affected blastocysts will be to be destroyed.

Premature ovarian failure: Ovarian failure (or more correctly insufficiency) occurring earlier than expected, and before the age of 40. Generally, the cause is idiopathic and rarely is it due to autoimmune or genetic causes, though the latter are worth looking for to effectively warn any female relatives who might want to store oocytes or ovarian tissue. The Fragile X premutation is one such cause which I have seen occur in families, and those warned of the diagnosis can consider some form of fertility preservation.

Progesterone: A sex steroid, produced in the luteinised granulosa cells of the corpus luteum, this is an important hormone in the stabilisation of the endometrium and bringing a halt to growth, promoting differentiation of glandular cells into the production of mucin and glycogen. At the end of the menstrual cycle, falling levels of progesterone bring about the shedding of the endometrium in an orderly manner. Crucial to the maintenance of an early pregnancy before the placenta begins to function.

Prolactin: One of the anterior pituitary hormones, prolactin is associated with milk production and in preparation of the breast to lactate, and negatively controlled by dopamine.

Pseudopuberty: The early onset of somatic and functional changes typical of puberty; commonly caused by endocrine secretions of an ovarian, testicular or adrenocortical tumour and typically arising before the chronologic age of puberty, and *not* related to the early normal initiation of puberty by hypothalamic-pituitary gonadotrophins.

Pubarche: As part of normal sexual development, this is the first appearance of pubic hair at puberty.

RCT: A study design set up to examine one treatment against another, or against a control. This is regarded as being the strongest model of study design, when numbers to achieve significance are set in advance, the patients and study members are often blind to what treatment is being used and patients are allocated to treatment or control by a process of randomisation.

recFSH: An artificial form of FSH, made by recombinant technology, and therefore not subject to supply chain problems or problems from from urinary contaminants in the way that urinary FSH can be.

Retrograde ejaculation: Passage of sperm at ejaculation into the bladder (in a backwards/retrograde manner). It is caused by inadequate tightening of the bladder neck muscles, and that is generally a result of a postsurgical damage or from nerve damage from medical causes (diabetes, a spinal cord injury) or an unwanted side effect of medication such as some antihypertensives (alpha blockers in particular).

Rokitansky's syndrome: See Mayer–Rokitansky–Küster–Hauser syndrome.

Rubella, or German measles: German measles, a generally mild illness in a woman. If it develops in the first half of pregnancy, it can have serious consequences for the baby, possibly causing blindness or deafness.

Schistosomiasis: A parasitic infection found in the tropics, usually in Africa, caused by flatworms called schistosomes. The urinary tract or the intestines may be infected. This is a rare cause of Asherman's syndrome.

Secondary amenorrhoea: Any cessation of periods once they have stopped – the most common cause is pregnancy, which should always be suspected when it occurs. The second most common cause in the developed world is probably polycystic ovary syndrome.

Selective salpingography: A radiological investigation using radio-opaque dye outlining the uterus and tubes and selective guidewire catheterisation of the fallopian tube to unblock it.

Semen: The carrier fluid, including sperm (no more than 1%) as well as fructose, magnesium, potassium, sodium and zinc, released by a man at ejaculation.

Seminal fluid: See Semen.

Seminal fluid analysis: A basic investigation for sperm, which examines numbers, motility and normality. It tells almost nothing about sperm quality or function.

Seminal vesicles: The organs in which seminal fluid is added to sperm to help carry the sperm into the vagina and cervix; they contain almost no sperm.

Seminiferous tubules: Coiled threadlike tubules which make up the bulk of the testis; the tubules are lined with epithelial cells from which the spermatozoa are produced. There is a distinct appearance at open surgery for tubules containing cells producing sperm (brown in colour) compared with those not doing so (pale yellow).

Sertoli cells: Somatic cells of the testis, essential for testis formation and spermatogenesis. They support germ cell progression to spermatozoa through both direct contact and controlling the environment milieu within the seminiferous tubules. FSH and testosterone control the regulation of spermatogenesis through their hormones working on the Sertoli cells.

Sexually transmitted infection (STI): A number of infections which are generally but not exclusively spread by sexual contact – including *Neisseria gonorrhoeae*, *Chlamydia trachomatis*, *Treponema pallidum* (syphilis) and others. They can have symptoms, and often their effect is not seen until much later in life (especially Chlamydia and gonorrhoea). Syphilis and Chlamydia can also be spread across the placenta in pregnancy and through the birth canal at delivery.

Sildenafil: A medication which inhibits phosphodiesterase 5, an enzyme that promotes degradation of cGMP, which regulates blood flow in the penis. Taking it leads to improved blood flow to the penis, thus improving the quality of an erection. There are now several phosphodiesterase 5 inhibitors on the market.

Sperm: Microscopic single-cell structure present in the fluid that a man ejaculates at orgasm. Sperm have the potential to penetrate and fertilise an oocyte to form an embryo. Produced by the male in the seminiferous tubules of the testis, they contain the haploid genetic content of the man.

Sperm disorder: See Sperm dysfunction.

Sperm dysfunction: The designed purpose of sperm is to fertilise oocytes. Any abnormality which disrupts that (in terms of numbers, normality or motility) will give rise to some degree of sperm disorder/dysfunction.

Sperm recovery and survival: A two-part test, designed to predict the likelihood of fertilisation at IVF, the really fundamental test of sperm function. It combines a swim-up procedure with survival for 24 hours. A swim-up procedure segregates motile and immotile spermatozoon. The ability of an individual spermatozoon to "swim" through a column of culture media within a given time indicates its cell velocity. Survival over 24 hours indicates the ability of sperm to survive that long and even longer, but that 24-hour time period is critical in permitting oocyte penetration by a sperm.

Spermatogenesis: The whole developmental process of sperm production from immature spermatogonia to mature sperms being produced at ejaculation. Some of this is hormone dependent, and the process takes 90–120 days.

Spindles: Part of the cytoskeleton within the cell, important during mitosis to draw the chromosomes apart and in meiosis to draw apart the polar body from the haploid content of the cell, as it is about to undergo fertilisation. The spindle structures are still in place during the mitotic division that takes place after fertilisation into a zygote. They are very fragile and easily damaged by cell distortion (as happens in freezing).

SSR: Surgical sperm recovery, a broad term to cover all the various approaches to retrieving sperm by surgical means, generally under anaesthesia. This includes aspiration of the epididymis, needle aspiration of the testis and testicular biopsy to retrieve sperm-carrying tissue. Testicular tissue containing sperm has a distinctive appearance, so there can be application of science to the biopsy process. Any such invasive technique will cause internal testicular scarring and also cause leakage of sperm outside the blood-testis barrier, perhaps leading to antisperm antibody formation.

Surrogacy: See Surrogate motherhood.

Surrogate motherhood: A process whereby a woman agrees to become pregnant and carries a child on behalf of another woman (usually with a legal agreement in place), either from her own egg fertilised by the other woman's partner (traditional), or from the implantation in her uterus of a fertilised egg from the other woman (gestational).

Syphilis: A sexually transmitted infectious disease, caused by *Treponema pallidum*, a spirochaete, which presents generally first as primary syphilis, with a painless raised chancre, later as a rash (secondary syphilis) and finally with neurological disorders as tertiary syphilis. We screen for it, as congenital syphilis occurs in up to 50% of mothers with syphilis and the disease causes profound neurological damage, and in recent years, several cases of pregnancy related syphilis have been identified in our city's maternity hospitals. Treatment is generally with penicillin-like antibiotics.

Tanner staging: In 1962, James Tanner (a London paediatric endocrinologist) published his staging system for the genital (males), breast (females) and body hair (both sexes), describing and assigning the development of these markers of pubertal advancement to specific ages (Chapter 1, Figure 1.2).

Testis/testicle: One of two organs lying in the scrotum, which produces most of the androgens in the male and is the site of spermatogenesis once puberty begins, lasting until the age of 55 years or so. Their external environment means their ambient temperature is about 1°C lower than the normal body temperature.

Testosterone, 17OH-progesterone, dehydroepiandrosterone sulphate: Markers of congenital adrenal hyperplasia due to 21-hydroxylase deficiency (produced in the adrenal zona reticularis); used for diagnosis and also responsiveness to treatment. They are usually reserved for determining the cause of hyperandrogenism, expressed as hirsutism or other symptoms. Prior to diagnosis and treatment, levels will generally be high for chronological and bone age.

Thelarche: The first stage of pubertal development with the growth of the breast bud, usually being visible not long after the age of 8 years. On the Tanner classification, this is called Tanner Stage 2 (Stage 1 being essentially prepubertal and undeveloped).

Thyroid-stimulating hormone: A glycoprotein secreted in the pituitary gland, with an α and β chain. The α chain is common to the other pituitary glycoproteins (FSH, LH) and to hCG; the β chain gives TSH its specificity in action and structure: stimulation of thyroxine and tri-iodothyronine production from the thyroid gland.

Transphenoidal hypophysectomy: Removal of the pituitary gland, via the nose, and passing through the sphenoid bone. Usually reserved for tumours refractory to medical treatment, and with side effects consistent with pituitary failure, requiring long-term replacement therapy of all the pituitary hormones.

Transvaginal oocyte recovery: Now the standard method of collecting oocytes with a view to *in vitro* fertilisation (as illustrated in Chapter 8, Figure 8.1).

Tubal damage: Any internal or external structural change, usually caused by infection, which alters the ability of the fallopian tube to move to pick up an oocyte at ovulation or impairs the passage of the oocyte down the lumen of the tube to meet with sperm, or the fertilised oocyte (zygote) to the endometrial cavity.

Tuberculosis: An infectious disease caused by *Mycobacterium tuberculosis*, now rarely found in developing countries, but still present in many developing countries, and infection in women, when present in the endometrium, can lead to a low-grade inflammatory response which destroys the endometrium leading to Asherman's syndrome.

Ullipristal: A selective progesterone receptor modulator (SPRM), used in reproductive medicine to suppress endometrium and myometrium when treating fibroids (myomata).

Ultrasound: An imaging modality which uses sound waves, and the images are generated by the different sonic reflective properties of physical structures.

Unexplained infertility: A diagnosis in a couple in whom all the tests are normal (tests of sperm numbers and quality, ovulation and tubal function) and no cause has been found for their subfertility.

Urethra: The passageway between the bladder and the external opening for passage of urine in the vulva (for women) or on the prepuce (for men).

Uterus: A midline structure in the pelvis formed by the fusion of the Mullerian tracts; which contains the myometrium (the muscle structure) and the endometrium (the lining which sheds at menstruation and is the site of pregnancy implantation).

Vas deferens: A thin cordlike structure above the testes, leading from the epididymis to the male urethra, along which sperm pass through and finalise their maturation process.

Y micro-deletions: Sections of the Y chromosome are recognised to be faulty or absent; there are three potential absent sections on the Y chromosome which are called azoospermia factor (AZF) a, b and c. Each has differing effects in the amount of sperm production that is restricted – AZFc being the least worst of the three.

Zona pellucida: A glycoprotein layer surrounding the plasma membrane of oocytes, surrounded by the cumulus oophorus. Sperm bind to the zona pellucida before they penetrate the oocyte, and the zona pellucida has to hatch before the fertilised oocyte (64-cell blastocyst) can implant.

Zygote: The cell (with 44 + XX or 44 + XY chromosomes) formed by and after the fusion of the male gamete (sperm, with 22 + X/Y chromosomes) with the female gamete (oocyte, with 22 + X chromosomes).

Problems with puberty and its onset

1

OVERVIEW

This chapter presents an understanding of the controlling mechanisms for the onset of puberty, and for precocious and delayed puberty. It provides an insight into the pathology and symptomatology, the investigation and the management of disorders of puberty. Of importance to this book, those disorders of puberty which contribute to reproductive ability are highlighted.

The clinical problems of puberty can be challenging for many trainees (and consultants), because the conditions are rare and infrequently encountered. Before addressing the causation and the investigation of the condition, it helps to understand the extent of the problem and to break down investigation into logical steps.

DEFINITIONS

Early or precocious puberty is indicated when, on Tanner staging (see Figures 1.1 and 1.2), girls reach the second Tanner breast stage before 8 years, and boys reach the second Tanner gonadal stage before the age of 9 years.

Central or true precocious puberty is *gonadotrophin dependent*, with increased activity in the arcuate nucleus of the hypothalamus. Central, true precocious puberty is divided into organic central precocious puberty, which has a discernible physical cause, and idiopathic central precocious puberty, which has no obvious cause.

Pseudoprecocious puberty is *gonadotrophin independent*, with peripheral causes, either endogenous or exogenous.

Delayed puberty (12) is recognised in females when breast Tanner stage 2 occurs after the age of 13; in males, delayed puberty is when testicular enlargement (beyond 4 mLs) is absent at the age of 14 years (UK population data) (5,6).

NUMBERS

In the United Kingdom, in the Avon Longitudinal Study of Parents and Children (ALSPAC), data were collected from children recruited at birth in 1991 and 1992, who have been followed up ever since (1,2). Progress through puberty was assessed using self-report questionnaires. For girls, 12% reported achieving Tanner breast stage 2 at age 8, while 98% reported being at stage 2 by age 13. For pubic hair, a Tanner stage 2 was reported by 4% and 95% of girls at ages 8 and 13, respectively. One girl (of 2953) reported having menarche by age 8; 60% had their menarche by age 13 (1). In this study,

- Only 2% of girls did not have secondary sex characteristics by the age of 13, but 40% of girls in their 13th year had not started menstruating.
- 5%–12% of girls reported secondary sexual characteristics by age 8 (suggesting precocious puberty), although only one child had a period by the same age.
- Pubic hair develops later than breast growth.

The ALSPAC report on boys presented only data on pubic hair growth. For boys, 5% reported Tanner pubic hair stage 2 at age 8; 99% were at stage 2 by age 14 (2).

When 9000 referrals to an Australian paediatric endocrinology clinic were reviewed, pubertal disorders were seen in 12% of cases (n = 1092) (4). Girls accounted for 715 of these, boys for 377. There was a marked gender difference in the types of presentation: 43% of the boys presented with

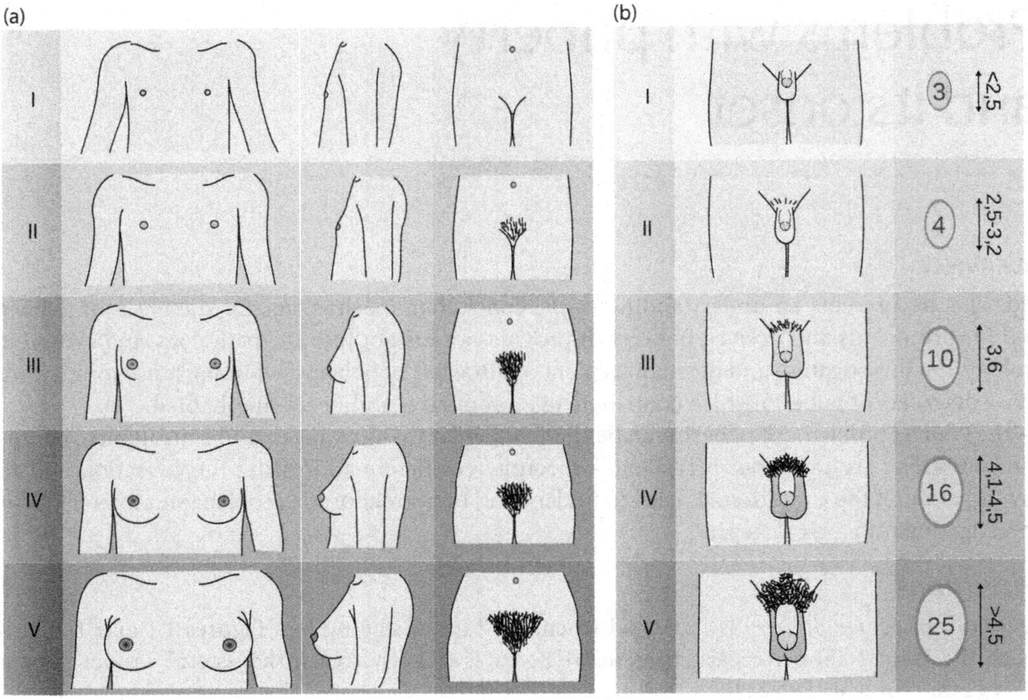

Figure 1.1 Pictorial representation of Tanner staging in (a) female and (b) male adolescents. (From Wikimedia Commons, http://goo.gl/haB9Cb, and http://goo.gl/7cxTLM under Creative Commons.)

delayed puberty versus 13% of the girls, whereas 28% of the boys presented with early onset of puberty compared with 75% of the girls.

Of those referred with suspected early/precocious puberty (n = 639), only half met the definition criteria for this diagnosis (n = 334). More girls than boys were discovered to have central precocious puberty, meaning gonadotrophin dependent (247 vs 63), but the actual reason was detected more in boys than girls, 40% vs 23%. A clear causative disease was found in 106 cases. These causes were predominantly hydrocephalus, brain tumours and mental retardation. Peripheral precocious puberty (not gonadotrophin dependent) was most commonly caused by non-central nervous system (CNS) tumours releasing sex hormones and follicular ovarian cysts (4). In another series of 213 children presenting with precocious puberty, Bridges and colleagues (7) found 91 girls and 4 boys with central precocious puberty; all the boys were found to have a cause, but only 6 of the 91 girls. *In summary, it is often not possible to find a cause for precocious puberty in girls, early puberty in boys usually has a detectable reason and many more children are referred for possible precocious puberty than have it.*

The Australian review found that delayed puberty as a variation of normal – termed constitutional or familial delay – occurred in 75% of cases. In the remaining 25%, females had an increased incidence of organic causes compared to males, related to gonadal dysgenesis, an eating disorder or the excessive thinness of elite athletes and dancers. Other organic causes (chronic disease, hypopituitarism or gonadal damage from radiotherapy) were similar in both sexes.

An American cohort of 859 children, recruited from 6 months old and followed up to the age of 15.5 years, was assessed for pubertal development by nurse practitioners (8). The study found no examples of girls or boys with breast or gonadal development before the ages of 8 and 9, respectively, that is, no precocious puberty. All children had entered puberty (Tanner stage 2) by the age of 13 for girls and by the age of 14 for boys. No child in their (admittedly small) cohort had either precocious or delayed puberty.

(a)

Stage	Breast growth	Pubic hair growth	Other changes	Age range (Year)
I	Pre-adolescent	None	Pre-adolescent	0–15
II	Breast budding (thelarche); areolar hyperplasia with small amount of breast tissue	Long, downy pubic hair near the labia, often appearing with breast budding or several weeks or months later	Peak growth velocity often occurs soon after stage II	8 or 8½–15
III	Further enlargement of breast tissue and areola, with no separation of their contours	Increase in amount and pigmentation of hair	Menarche occurs in 25% of girls late in stage III	10–15
IV	Separation of contour; areola and nipple form secondary mound above breast tissue	Adult in type but not in distribution	Menarche occurs in most girls in stage IV, 1–3 years after thelarche	10–17
V	Large breast with single contour	Adult in distribution	Menarche occurs in 10% of girls in stage V	12½–18

(b)

Stage	Testes growth	Penis growth	Pubic hair growth	Other changes	Age range (Year)
I	Pre-adolescent testes ≤2.5 cm	Pre-adolescent	None	Pre-adolescent	0–15
II	Enlargement of testes; pigmentation of scrotal sac	Minimal or no enlargement	Long, downy hair, often appearing several months after testicular growth; variable pattern noted with pubarche	–	10–15
III	Further enlargement	Significant enlargement, especially in diameter	Increase in amount; curling	–	10½–16½
IV	Further enlargement	Further enlargement, especially in diameter	Adult in type but not in distribution	Axillary hair and some facial hair develop	Variable (12–17)
V	Adult in size	Adult in size	Adult in distribution (medial aspects of thighs; linea alba)	Body hair continues to grow and muscles continue to increase in size for several months to years; 20% of boys reach peak growth velocity	13-18

Figure 1.2 (a) Sexual maturity rating (Tanner staging) in female adolescents. (b) Sexual maturity rating (Tanner staging) in male adolescents: values refer to testicular volume in mLs and length of the testis in cm. (From Tindyebwa D. *Handbook on Pædiatric AIDS in Africa*. 2nd ed., Vol. 1. Kampala, Uganda: African Network for the Care of Children Affected by AIDS; 2006, under Open Access.)

One further study from Europe (Denmark) examined the Danish national population registry for the diagnosis of precocious puberty over an 8-year period (9). They found that 50–70 new cases presented each year, with a prevalence rate of 0.2% in Danish girls and <0.05% in Danish boys.

CONCLUSION

Drawing on the four studies above, one can draw the following conclusions. The concerns of parents and care givers, and sometimes the young person, result in referrals to specialist centres for both precocious and delayed puberty, but only a proportion, roughly half, meet the criteria for diagnosis for precocious puberty. Reassurance and explanation are important in the cases not meeting the criteria for diagnosis. Self-assessment results in higher findings of pubic hair and genital development than assessment carried out by trained personnel. Precise figures on the incidence and prevalence of precocious puberty are difficult to obtain. Population figures appear to be low – 0.2% prevalence in Danish girls and much, much lower in boys. However, it is important to detect precocious and delayed puberty, for the reasons explained in the relevant sections below.

1.1 HOW DOES PUBERTY HAPPEN?

Puberty is a physical and psychological process that girls and boys experience as they develop sexual organs so that they can procreate. Puberty is concluded by the completion of gonadal and genital development, breast and pubic hair maturity and the attainment of final height. Problems related to puberty are that it presents itself too early (early or precocious) or too late (delayed) – summarised in Figure 1.3.

The first physical evidence of puberty in girls is a growth spurt. This is followed by the development of breast tissue (thelarche), the appearance of pubic and axillary hair (adrenarche; a different word is used for hair growth in girls and boys because the androgen source is different) and, finally, the onset of periods, which is menarche. Hair and breast development typically begin around age 9. The growth spurt starts around the age of 10, reaching a maximum velocity around the age of 13, and finishing by age 15, long after all the other outward signs of puberty are completed (10). Not visible is the growth of the vagina, uterus and ovaries, which occurs alongside the growth of breast tissue. Only by imaging techniques can these organs be assessed, and there are some published data available to undertake this assessment (13). All that can be seen are the labia, vulva and pubic hair – and the lack of reliable material as a comparator can often lead to great anxiety for young women and their parents (11). There is no special term for the process of vagina and uterus development.

In boys, the first physical evidence of puberty is the growth in testicular volume and penis size (gonadarche). The growth spurt in boys starts after genital and gonadal growth begins. This is

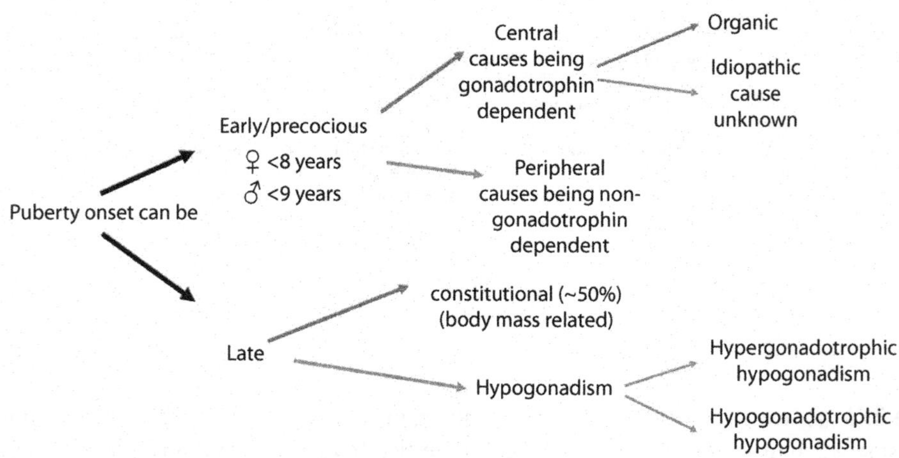

Figure 1.3 How puberty problems present.

followed by the appearance of pubic and axillary hair (called pubarche in boys) (10). The growth spurt generally is completed by the age of 16, long after all the other signs of puberty are complete.

1.1.1 What initiates puberty?

It is not known what initiates puberty, and this is an important area of research and enquiry. There is evidence to suggest that, in the prepubertal child, the immature gonads release an inhibitory chemical mediator which suppresses the release of gonadotrophin-releasing hormone (GnRH) by the hypothalamus (12). In normal boys with immature gonads, low GnRH levels during childhood have been found, but high levels of GnRH (and follicle-stimulating hormone and luteinising hormone – FSH and LH) were found in boys without testes (10,13). This has given rise to the theory of a 'brake' operating on the hypothalamus, preventing GnRH release. What is not understood is what releases the brake when the child is ready to commence puberty. The brake is evidently more profound and longer lasting in boys than it is in girls, demonstrated by the earlier onset of puberty in girls and the higher proportions of girls having 'precocious' puberty (see Section 1.2). As the brake is released, GnRH pulses begin, initially at night, and these induce gonadotrophin secretion, which brings about gonadal stimulation (FSH in girls to produce oestrogen; LH in boys to produce testosterone).

While the trigger for the onset of puberty is still unclear, there is increasing knowledge about the peptides that control the secretion of GnRH by the arcuate nucleus (12). A group of neurones, the anteroventral periventricular nucleus – sited close to the lateral walls of the third ventricle in the preoptic area – and neurones in the arcuate nucleus release kisspeptin, a neuropeptide hormone, so named because it was first identified in Hershey, Pennsylvania, USA, the home of Hershey's Kisses (14). Receptors for kisspeptin (KISS1R), found in the anteroventral periventricular nucleus, seem to be as important as the hormone itself. Absence of the gene for the receptor has been linked with idiopathic hypogonadotrophic hypogonadism (15). Gamma-aminobutyric acid (GABA – suppressive) and glutamate (stimulatory) are also active at the anteroventral periventricular nucleus level, while at the arcuate nucleus, kisspeptin neurones also secrete neurokinin B and dynorphin A (Figure 1.4). Before puberty starts, dynorphin A has a dominant inhibitory influence. As puberty starts, neurokinin B increases in its stimulatory drive overcoming dynorphin A, allowing kisspeptin to become active and stimulatory as well (12). From this, it seems reasonable to speculate that whatever increases neurokinin B is the responsible factor for the initiation of the GnRH pulses.

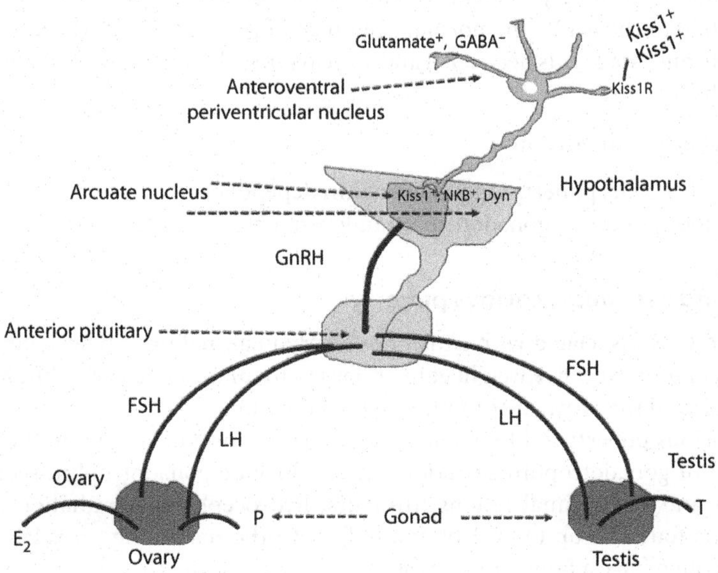

Figure 1.4 GnRH activation in puberty.

The operation of the hypothalamic pituitary adrenal (HPA) axis in girls, releasing androgen to stimulate pubic and axillary hair growth, is puzzling for two reasons. First, it's not clear what purpose androgen release and hair development serve in girls and why the body retains this feature. Second, the mechanism which triggers the release of androgens at the start of puberty is completely unknown. There are various theories, involving POMC – proopiomelanocortin – found in the anterior pituitary, or the zona reticularis in the adrenal gland, or phosphorylation of P450c17, necessary for 17,20-lyase activity. None of the theories has much supporting evidence, and the mechanism remains a mystery (16).

1.1.2 Does the onset of puberty vary between different populations?

Puberty onset appears not to vary between ethnic populations – diet and socioeconomic factors appear to exact more influence. Two large cross-sectional data studies in white and black American girls did show that there were differences in the onset of breast development (Tanner stage B2), but almost no difference in the onset of periods (menarche). Breast development was self-reported in these studies and it may be that fat deposition was mistaken for glandular breast development (17). In developing countries, where differences in diet and socioeconomic status are greater, disparity in menarche is seen; girls from less privileged backgrounds have menarche relatively later in life, around 16 years (17). Changes in the timing of the onset of menarche over the centuries are well known. From the 1850s to the 1950s, the average age of menarche in developed North America and Europe fell from 17 to under 14 years (18), reinforcing the link with economic development and status.

1.2 PRECOCIOUS PUBERTY

There are two problems seen when puberty starts too early:

- The *psychological distress* caused to child and family when physical appearance and periods are out of step with chronological age.
- The *negative impact on eventual growth* because precocious puberty leads to accelerated growth and bone maturation and unwanted epiphyseal closure, and ultimately reduced stature (this can be as much as 20 cm for boys and 12 cm for girls) (19).

Precocious puberty is more prevalent in females than males, 20 per 10,000 in girls compared with <5 per 10,000 for boys (9). The normal age range of pubertal onset, the age at which 95% of children attain Tanner stage 2, is between 8 and 13 years in girls, and between 9 years 6 months and 13 years 6 months in boys.

Precocious puberty is divided into

- Central/true precocious puberty – gonadotrophin dependent
- Pseudoprecocious puberty – gonadotrophin independent

1.2.1 Central and pseudoprecocious puberty

Early onset puberty is associated with a history of early maternal menarche, a personal history of low birth weight and/or excessive weight gain or obesity in infancy and early childhood, after international adoption and the absence of a father in the house (3).

Central precocious puberty (CPP) is caused by the early onset of activity in the arcuate nucleus with early onset of gonadotrophin secretion. Causes include gonadotrophin-secreting tumours, hypothyroidism and intracranial lesions (tumours, hydrocephalus, irradiation and trauma). A definitive cause is found in up to 40% of cases (4), with the remaining cases labelled idiopathic. Idiopathic cases often have a family history of early puberty, or the child is overweight or obese. In girls, idiopathic is the most common diagnosis in gonadotrophin-dependent early puberty, while a definitive cause is usually found in boys.

Pseudoprecocious puberty is diagnosed when there is no evidence for early gonadotrophin secretion from the pituitary. The physical signs of puberty are instead due to 'peripheral' causes. These include congenital adrenal hyperplasia, sex steroid secreting tumours (ovarian or adrenal), McCune-Albright syndrome (due to ovarian or testicular benign but endocrinologically active tumours) and accidental ingestion of oestrogen (perhaps as the oral contraceptive pill). Congenital adrenal hyperplasia is the most common condition found.

1.2.2 Assessment and investigations

When a child presents to a primary or secondary care setting, examination of the breasts and/or genitals should not be the first step (Figure 1.5). Get a full personal and family (of other siblings) history, and measure height and weight, plotting it onto a growth chart. Assessment of the breasts/genitals can be done by physical examination or by agreement using a Tanner chart, depending on the child. It may be that on this first assessment, the findings are inconclusive – it makes sense to wait 3–6 months to re-evaluate before moving on to the full investigation. Once the signs of precocious puberty are evident, begin the investigations as outlined below (Figure 1.5).

A step-by-step approach to assessing early puberty includes:

Step 1: Bone age measurement: wrist small bone appearance should be correlated with chronological age. Disparities highlight early or late development. Bone age in precocious puberty is usually more advanced than chronological age.

Step 2: Measurement of hormones (LH, human chorionic gonadotrophin [hCG], oestradiol, testosterone, 17OH-progesterone and DHEAS – dehydroepiandrosterone sulphate). LH should be measured in the early morning or following a GnRH stimulus. Any LH is indicative of precocious puberty. Elevated testosterone is indicative of precocious puberty (greater than 100 ng/dl or 4 nmol/L), while oestradiol is more variable and not very sensitive for the diagnosis. Levels of >24 pg/mL or >80 pmol/L are suggestive, but most laboratories will not report values less than 27 pg/mL or 100 pmol/L. If hCG is present, it indicates a germ cell tumour in the ovary or testis. 17OH-progesterone and DHEAS elevation is indicative of congenital adrenal hyperplasia (CAH).

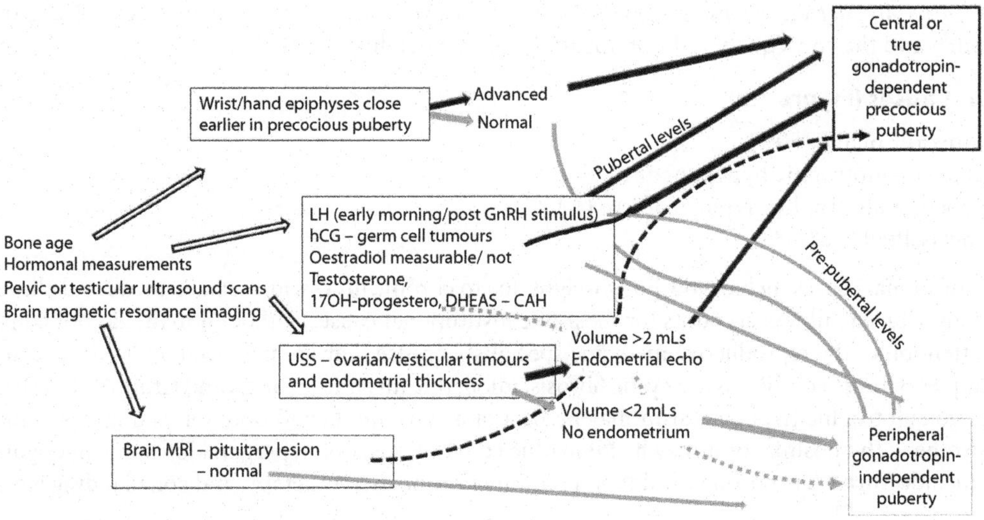

Figure 1.5 Step-by-step approach to assessing early puberty: central true puberty is noted by black (solid or broken) lines; pseudoprecocious puberty or normality is noted by grey (solid or broken) lines.

Step 3: Ultrasound images of the pelvis or testicles (assessing uterine/testicular size). For boys, testicular ultrasound is indicated if testicular volume is asymmetric or if peripheral precocious puberty is present (to rule out a Leydig-cell tumour). In girls with peripheral precocious puberty, ultrasound may demonstrate a hormonally active tumour. Uterine volume increases (>2 mLs) are a very sensitive and specific test for precocious puberty diagnosis. Refer to Tanner charts for normal values.

Step 4: Magnetic resonance imaging (MRI) imaging of the brain (focussing on the hypothalamus and pituitary) should be performed in all boys with CPP, girls with CPP with onset before age 6 and any girl with clinical findings indicating the possibility of a CNS lesion (e.g. severe frequent headaches, changes in vision or new-onset seizures) to exclude a hypothalamic hamartoma (3,20). Managing girls with CPP between 6 to 8 years old requires discussion with the parents and child about the relative risks and benefits of the procedure (20).

1.2.3 Management of early puberty

Management of central precocious puberty is by suppression of LH and FSH, using depot injections of a gonadotrophin-releasing hormone agonist (GnRHa). Treatment continues until the female child is judged old enough, by parents and doctor, to have periods, ideally by age 11. There is a variable delay in onset of puberty once the GnRH agonist is stopped. No adverse effects are apparent on reproductive potential. The effect of treatment on final height is not so successful (19). Treatment with GnRH agonist should be started as soon as possible after diagnosis; early treatment is advisable because of bone maturation (19).

For pseudoprecocious puberty with gonadotrophin-independent causes, treatment is for the offending cause, whatever that might be: treatment of congenital adrenal hyperplasia, surgery for sex steroid–secreting tumours and aromatase inhibitors (letrozole and anastrozole) for McCune-Albright syndrome (21).

1.3 DELAYED PUBERTY

There are two problems seen when puberty is delayed

- *Psychological distress* (for girls, the absence of breast development and the absence of periods; for boys, the lack of genital development and short stature)
- *Negative impact on eventual growth potential* (delayed puberty leading to a lack of a growth spurt, and the consequent risks of not attaining full height potential)

1.3.1 Causes (Figure 1.6)

- Constitutional (~50%)
- Hypogonadotrophic hypogonadism (~33%)
- Hypergonadotrophic hypogonadism (~15%)
- Unclassified (~2%–3%)

Constitutional causes include low body weight due to chronic illness, anorexia nervosa and elite athleticism. Chronic illness accounts for most of constitutional causes, although many do not come to the attention of the reproductive medicine specialist because they are managed by other specialists. Examples of these conditions are cystic fibrosis, idiopathic juvenile arthritis and Crohn's disease (26). Hypogonadotrophic hypogonadism may be part of a congenital condition, for example, Kallmann syndrome with anosmia, or it may be idiopathic (IHH). Causes of hypergonadotrophic hypogonadism include Turner's syndrome, childhood radiotherapy or chemotherapy, and gonadal dysgenesis.

1.3.2 Assessment and investigations

Assessment includes history taking, covering general physical and psychological health, looking for clues suggesting a chronic physical condition and family history of pubertal onset. This is followed by physical examination, height and weight measurement and pubertal staging by direct

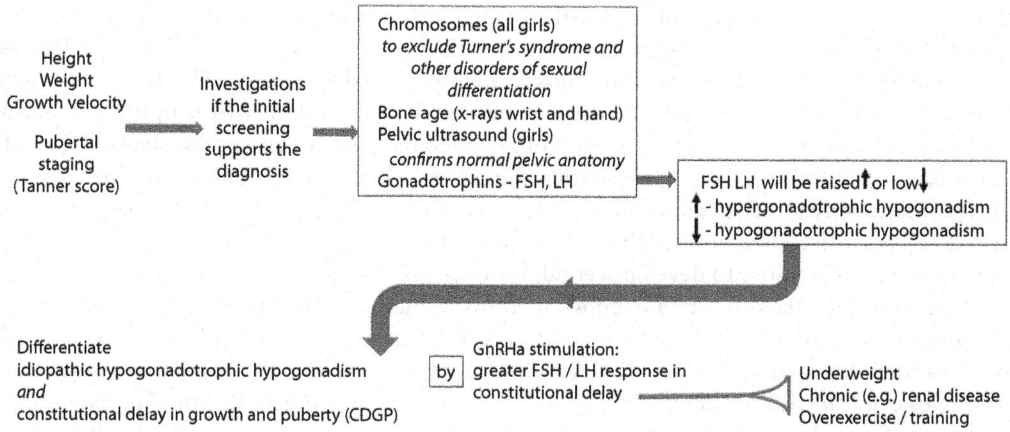

Figure 1.6 Step-by-step approach to assessing delayed puberty (read in conjunction with Section 1.3).

observation or using pictorial representations of Tanner staging. It is helpful to bear in mind the findings of the ALSPAC study (2) showing that in boys, self-assessment is not as accurate as physical examination, with younger boys tending to overestimate their development.

For girls, the absence of periods and secondary sexual characteristics by the age of 14 should be investigated. If amenorrhoeic but exhibiting secondary characteristics by the age of 16, the most likely cause is a Mullerian structural anomaly causing a blockage of the outflow tract, such as an imperforate hymen or a more extensive occlusion, or agenesis such as Mayer–Rokitansky–Küster–Hauser syndrome with failed development of the vagina, cervix and uterus. The latter condition is very rare. In Bristol, there are about 10,000 deliveries a year of which 5000 are females, and 1 case of Rokitansky's syndrome is likely to occur in that number. The presumptive diagnosis of an imperforate hymen should not be approached surgically without being completely sure of the state of the vagina (above the hymen, most especially, is there a vagina?). MRI is probably the most useful way to assess this.

In the investigation of delayed puberty in boys, the differential diagnosis is usually idiopathic hypogonadotropic hypogonadism (IHH) or constitutional delay in growth and puberty (CDGP). Confirmation is complicated by the lack of a simple biochemical test to differentiate between CDGP and IHH. GnRHa stimulation is reputed to be effective in discriminating between the two conditions (12). Basal inhibin B is being investigated as a diagnostic test, but the evidence is, as yet, inconclusive (22).

Investigations should follow the steps outlined in Figure 1.6.

Step 1: Blood testing. Full blood count, thyroid function tests, liver and renal function tests to establish basic well-being. Karyotype testing for Turner's syndrome if physical examination suggests this possibility.

Step 2: Bone age is usually delayed in late onset puberty.

Step 3: Ultrasound will confirm or refute the presence of normal anatomy in the pelvis.

Step 4: Serum gonadotrophins levels are useful if raised as this is diagnostic of hypergonadotrophic hypogonadism. Low levels need further assessment using GnRH stimulation. A good response (increased levels of FSH and LH above 2 IU/L) indicates constitutional delay as causal, whereas a blunted response (<1 IU/L – as an indicator) occurs in hypogonadotropic hypogonadism (12).

1.3.3 Management to induce puberty

Treatment of teenage boys and girls involves the use of testosterone and oestradiol, respectively. Oral testosterone has variable absorption and is broken down in the liver, so transdermal and injectable forms of natural or synthetic carboxylated testosterones are used. For girls, oral ethinyloestradiol or conjugated oestrogens are given, by oral, transdermal or injectable routes. Detailed descriptions of the modes of delivery and dosages are provided elsewhere (23). The key principle is to provide low and incremental treatment to ensure that skeletal growth is optimised but not blocked by early epiphyseal closure as a result of too rapid increases in exogenous steroids. This also ensures that secondary sexual characteristics, particularly breast development, are not adversely affected, since overly rapid increases in oestrogens or over-early introduction of progestogens can give rise to breast development which is asymmetrical or irregular. It may be necessary to resist patient or parental pressure to develop secondary sexual characteristics too rapidly.

In male teenagers, testosterone injections or transdermal gels administered over 3–4 weeks in graduated doses should bring about desired growth spurts and virilisation. If the cause of delayed puberty is constitutional, one course of treatment is usually sufficient, while IHH will require several courses until the desired height, penile and testicular size is reached. It should be explained to the patient with IHH that that testosterone therapy will suppress spermatogenesis, but that this can be reversed later with gonadotrophins, as explained below.

1.4 DISORDERS OF PUBERTY CAUSING REPRODUCTIVE PROBLEMS

There are a few pubertal disorders which have a major, sometimes irreversible, impact on the reproductive ability of the individual. Most of these conditions are present since birth but only become symptomatic around the time of puberty.

1.4.1 Girls

Conditions giving rise to reproductive problems are anatomical and endocrine in nature. Anatomical causes occur when there is failure of ovarian or uterine development, girls with Turner's syndrome, which occurs in 1 in 2500 live female births lack ovarian tissue. The girl will have characteristic physical features of the syndrome, but the diagnosis may not be made until delayed puberty is investigated. Other than Turner's syndrome, failure of ovarian development can occur for no apparent reason. Indicative external signs are the presence of axillary and pubic hair, but with no breast development. Other physical causes include those already mentioned, and the very rare, Mayer–Rokitansky–Küster–Hauser syndrome, with agenesis of the vagina, cervix and uterus. Ultrasonography of the pelvis and reproductive organs is needed to detect these conditions, while chromosome analysis is necessary to confirm Turner's syndrome. Because they possess a normal uterus, (but with almost no ovarian tissue), women with Turner's syndrome may be able to achieve a pregnancy using donated oocytes. However they will require a comprehensive assessment of their cardiovascular system, as life-threatening cardiac conditions, such as aortic dissection, have been reported in pregnancy (24). In those women who anatomically lack both ovaries and uterus, sadly little can be done to achieve a pregnancy.

Endocrine problems are more amenable to medical management. Hypogonadotrophic hypogonadism, either idiopathic or Kallmann's syndrome, is amenable to treatment. In these conditions, failure of the hypothalamic arcuate nucleus to secrete GnRH means there is no stimulation of the gonadotroph cells in the anterior pituitary resulting in an unstimulated ovary, and failure to produce oestradiol. As a consequence, oestrogen-dependent external sexual characteristics do not occur. Girls with hypogonadotrophic hypogonadism require a medically induced puberty with the administration of oestrogen, as outlined above. When the time comes to attempt a pregnancy, they will need ovulation induction with gonadotrophins as described in Chapter 5.

1.4.2 Boys

Pubertal problems affecting reproduction in boys are endocrine in nature, caused by hypogonadotrophic hypogonadism. Treatment uses testosterone replacement, as described above. Induction of spermatogenesis using gonadotrophins is described in Chapter 4 (Section 4.6.1). Success with this treatment is encouraging. In our experience, 12 out of 16 men produced sufficient sperm after treatment to enable some to be frozen, and 9 conceived a child (25).

1.5 SUMMARY

Scoring for the stages of puberty uses the Tanner classification, based on external features only. Early/precocious puberty is detected in girls when they reach the second Tanner stage for breast development before 8 years. In boys, it is detected when they reach the second Tanner stage for gonadal and genital growth before the age of 9 years. Central/true precocious puberty is gonadotrophin dependent; pseudoprecocious puberty is gonadotrophin independent. Delayed puberty in girls is recognised when puberty has not occurred by the age of 13 years. In boys, it is recognised when testicular enlargement beyond 4 mLs is absent at the age of 14 years. Delayed puberty and short stature occur more often in boys than girls. Conversely, early puberty and tall stature occur more often in girls. The breakdown of referrals for pubertal disorders is 30% delayed onset, and 70% early onset, but of this 70%, only half have true precocious puberty.

REFERENCES

1. Rubin C, Maisonet M, Kieszak S, Monteilh C, Holmes A, Flanders D et al. Timing of maturation and predictors of menarche in girls enrolled in a contemporary British cohort. *Paediatric and Perinatal Epidemiology.* 2009;23(5):492–504.
2. Monteilh C, Kieszak S, Flanders WD, Maisonet M, Rubin C, Holmes AK et al. Timing of maturation and predictors of Tanner stage transitions in boys enrolled in a contemporary British cohort. *Paediatric and Perinatal Epidemiology.* 2011;25(1):75–87.
3. Carel JC, Leger J. Clinical practice. Precocious puberty. *New England Journal of Medicine.* 2008;358(22):2366–77.
4. Thomsett MJ. The spectrum of clinical paediatric endocrinology: 28 years of referrals to an individual consultant. *Journal of Paediatrics and Child Health.* 2010;46(6):304–9.
5. Revised 2003 consensus on diagnostic criteria and long-term health risks related to polycystic ovary syndrome. *Fertility and Sterility.* 2004;81(1):19–25.
6. Christensen KY, Maisonet M, Rubin C, Flanders WD, Drews-Botsch C, Dominguez C et al. Characterization of the correlation between ages at entry into breast and pubic hair development. *Annals of Epidemiology.* 2010;20(5):405–8.
7. Bridges NA, Christopher JA, Hindmarsh PC, Brook CG. Sexual precocity: Sex incidence and aetiology. *Archives of Disease in Childhood.* 1994;70(2):116–8.
8. Susman EJ, Houts RM, Steinberg L, Belsky J, Cauffman E, Dehart G et al. Longitudinal development of secondary sexual characteristics in girls and boys between ages 91/2 and 151/2 years. *Archives of Pediatrics & Adolescent Medicine.* 2010;164(2):166–73.
9. Teilmann G, Pedersen CB, Jensen TK, Skakkebaek NE, Juul A. Prevalence and incidence of precocious pubertal development in Denmark: An epidemiologic study based on national registries. *Pediatrics.* 2005;116(6):1323–8.
10. Sanfilippo J, Jamieson M. Physiology of Puberty. Glob libr women's med [Internet]. 2008; doi:10.3843/GLOWM.10286. Available from: https://www.glowm.com/section_view/heading/Physiology%20of%20Puberty/item/285
11. Lloyd J, Crouch NS, Minto CL, Liao LM, Creighton SM. Female genital appearance: 'Normality' unfolds. *BJOG: An International Journal of Obstetrics and Gynaecology.* 2005;112(5):643–6.
12. Wei C, Crowne EC. Recent advances in the understanding and management of delayed puberty. *Archives of Disease in Childhood.* 2016;101(5):481–8.
13. Grinspon RP, Ropelato MG, Bedecarras P, Loreti N, Ballerini MG, Gottlieb S et al. Gonadotrophin secretion pattern in anorchid boys from birth to pubertal age: Pathophysiological aspects and diagnostic usefulness. *Clinical Endocrinology.* 2012;76(5):698–705.

14. Roseweir AK, Millar RP. The role of kisspeptin in the control of gonadotrophin secretion. *Human Reproduction Update.* 2009;15(2):203–12.
15. de Roux N, Genin E, Carel JC, Matsuda F, Chaussain JL, Milgrom E. Hypogonadotropic hypogonadism due to loss of function of the KiSS1-derived peptide receptor GPR54. *Proceedings of the National Academy of Sciences of the United States of America.* 2003;100(19):10972–6.
16. Fritz MA, Speroff L. *Clinical Gynecologic Endocrinology and Infertility.* Lippincott Williams & Wilkins; Philadelphia, Pennsylvania, 2010.
17. Parent A-S, Teilmann G, Juul A, Skakkebaek NE, Toppari J, Bourguignon J-P. The timing of normal puberty and the age limits of sexual precocity: Variations around the world, secular trends, and changes after migration. *Endocrine Reviews.* 2003;24(5):668–93.
18. Anderson SE, Must A. Interpreting the continued decline in the average age at menarche: results from two nationally representative surveys of U.S. girls studied 10 years apart. *Journal of Pediatrics.* 2005;147(6):753–60.
19. Mul D, Hughes IA. The use of GnRH agonists in precocious puberty. *European Journal of Endocrinology.* 2008;159(Suppl. 1):S3–8.
20. Kaplowitz PB. Do 6-8 year old girls with central precocious puberty need routine brain imaging? *International Journal of Pediatric Endocrinology.* 2016;2016:9.
21. Hickey M, Balen A. Menstrual disorders in adolescence: Investigation and management. *Human Reproduction Update.* 2003;9(5):493–504.
22. Harrington J, Palmert MR. Clinical review: Distinguishing constitutional delay of growth and puberty from isolated hypogonadotropic hypogonadism: Critical appraisal of available diagnostic tests. *Journal of Clinical Endocrinology and Metabolism.* 2012;97(9):3056–67.
23. Gynaecologists RCoOa. *Sex Steroid Treatment for Pubertal Induction and Replacement in the Adolescent Girl.* London: RCOG; 2013. p. 6.
24. Bouet PE, Godbout A, El Hachem H, Lefebvre M, Berube L, Dionne MD et al. Fertility and pregnancy in turner syndrome. *Journal of Obstetrics and Gynaecology Canada: JOGC = Journal D'obstetrique et Gynecologie du Canada: JOGC.* 2016;38(8):712–8.
25. Morris G, Cahill D. Gonadotrophin-induced spermatogenesis: Eight years' experience. *Human Fertility.* 2018;21(1):52–75.
26. Patel L. Growth and Chronic Disease. *Annales Nestlé (English ed).* 2007;65(3):129–36.

Understanding infertility* 2

OVERVIEW

This chapter covers the definition of infertility and its frequency, how different factors affect the likelihood of infertility, the diagnostic categories and whether the incidence of infertility is rising or falling. The available options for patients to achieve a pregnancy have expanded enormously over the last 30 years due to better surgical techniques and improved assisted conception techniques. This means that a number of fertility treatments are no longer experimental, with independently verified success rates. New, more experimental techniques are also available, but the drawback is that success rates are not widely available for those.

DEFINITIONS

Infertility means not being able to conceive. Very few couples are truly infertile, and will never be able to get pregnant. For most couples however, the problem is that they cannot get pregnant within a particular timeframe. The timeframe is commonly defined as 1 year of having regular vaginal intercourse, during which time 90% of couples having regular intercourse will get pregnant. In the United Kingdom, where health services are publicly funded, many health service organisations have extended the definition to 2 years, reflecting the fact that 95% of couples will have conceived by then. In this way, they reduce the numbers of couples being referred for investigation and for treatment. Inability to conceive within 2 years should be regarded as subfertility until absolute infertility is confirmed. Confusingly, some epidemiological studies use live births rather than ability to conceive as the defining feature of infertility.

PREVALENCE/INCIDENCE

Approximately 1 in 6–7 (14%–16%) couples in the United Kingdom is affected by substantial delay in conception (1,29). Similar findings are reported from other developed countries (10% of couples) (30) and developing countries (10.5%) (31). In developed countries, most couples presenting with delay will have primary infertility - never having conceived whilst in developing countries, most couple will have secondary infertility - having previously conceived. (See Section 10.1 in Chapter 10 for further details.)

2.1 FACTORS AFFECTING THE ABILITY TO CONCEIVE

2.1.1 Population differences

One population difference is seen in the incidence of *uterine fibroids*. Women whose racial origins are from the north coast of Africa and the coastline from Sierra Leone to Nigeria have a greater risk of myomata (fibroids) which can affect their fertility. Specific genetic markers for fibroids have been identified. In women of North African descent, these are seen on chromosome 6 in the p24 region (2).

Polycystic ovarian syndrome (PCOS) is a common cause of ovulatory dysfunction. In the UK population, up to a third of women have PCOS, but a higher prevalence has been noted in third-generation women from the South Asian peninsula (India, Sri Lanka), with more than 50% of such women living in the UK having the condition (3). Women whose ethnicity is Chinese or Japanese or broadly East Asian have a low prevalence of PCOS, of the order of 5%–6% (4).

Endometriosis is a clinical condition which in both its severe and its minor forms can affect fertility, and in the minor form, probably by altering follicular endocrinology, leading to a reduction

---

* Some material in this chapter is adapted from *Understanding Infertility*, Cahill DJ, Wardle PG. Family Doctor Publications, with permission.

in fertilisation rates of oocytes (5). Endometriosis itself has a particular racial distribution, being found in higher numbers in Japanese and other Asian women (6) and lower numbers in women from African countries (7) when compared with Caucasian women. These differences may be due to the way in which genetic and environmental factors interact. More for interest than any real scientific value, there are reports that show an association between having natural red hair and the development of endometriosis (8,9).

2.1.2 Age, body weight, frequency of intercourse

One of the most significant factors influencing infertility is *age*. Older women have considerably greater difficulty in conceiving than younger women do. Fertility is continuously declining in women, with a major increase in the rate of fertility loss from 37–38 years onwards. Malcolm Faddy and colleagues developed a model to demonstrate this. There is a steady loss of follicle numbers from birth onwards (10,11), but once the number of follicles falls to about 25,000 (usually around the age of 37.5 ± 1.2 years), the rate of follicle loss more than doubles. What this means in practice is that women have a steady loss in egg numbers up to the age of (say) 37 years, and thereafter that oocyte loss is doubled and time to loss of all oocytes is accelerated. Increasing age also affects fertility in men. The European Fecundability Study (14) found that men aged 40 had a lower chance of conception than men aged 35, with the same frequency of intercourse, and this finding has been replicated many times.

One further physical aspect deserves mention. Women who are *overweight* or *obese* are in a higher risk category for failure to conceive. A study group in Adelaide collected data on body mass index (BMI) (kg/m^2) in a cohort of women attempting pregnancy by *in vitro* fertilisation (IVF) (12). While BMI may not be the perfect measure of obesity, the data in Figure 2.1 shows an increasing risk of failure to conceive with IVF as BMI rises. Further data also demonstrates that with increasing BMI, the likelihood of miscarriage increases (13).

Frequency of intercourse has an impact on the likelihood of conception. Data from the European Fecundability Study (14) support the intuitive supposition that once a week intercourse is less likely to achieve a pregnancy than intercourse twice a week. Ejaculation up to eight times a week does not appear to diminish sperm numbers, but it does reduce sperm motility (15) and an interval of 3–4 days is recommended for a maximal total motile sperm count (16). Further evidence from treatment of couples with intrauterine insemination suggests that an interval of 2 days gives rise to optimal pregnancy rates (17). It would seem that the optimal frequency of sexual intercourse is every 2–3 days.

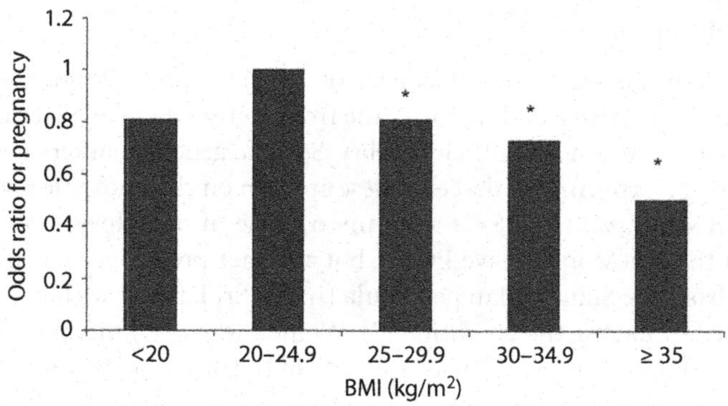

Figure 2.1 Likelihood of pregnancy following infertility treatment analysed by BMI (* indicates the change from the previous column is significant). (Adapted from data in Wang JX et al. *BMJ (Clinical Research Ed)*. 2000;321(7272):1320–1.)

2.2 DIAGNOSTIC CATEGORIES

The causes of infertility/reduced fertility can be grouped into broad diagnostic categories which will be examined in detail in the remainder of this book.

- Tubal damage
- Ovulatory disorder
- Sperm disorder
- Unexplained infertility
- Other (endometriosis, congenital or acquired anatomical defects)

2.3 BARRIERS TO SEEKING ADVICE?

Compared with most other mammals, humans have relatively poor natural fertility. In young fertile couples, the chance of conceiving in each monthly cycle is only 33% (1 in 3). This appears not to have changed despite people living longer. It is unchanged for the 60 years since data from the Hutterite population (who never used contraception) were published showing the same, monthly 1 in 3 chance of conception. (Figure 2.2) (18).

Some couples delay seeking medical advice because they find it difficult to accept that they have a problem or because they are concerned that they will be asked questions about personal and private aspects of their lives. Staff at clinics ought to be aware of these concerns. They should be prepared for the anxieties and fears about the investigations, the diagnostic problems that might be identified and the treatment options. The information that you as clinicians provide will enable patients to decide how they wish to proceed. Delay in seeking advice, or delay in making a decision may limit the treatment options and the chance of success for a couple.

Couples may be concerned that they will only be offered expensive, high-technology infertility treatments, such as IVF. A report from a workshop in 2015 did, in fact, propose that it might be more cost effective to simply treat all couples with 3 years of infertility (or longer) with *in vitro* fertilisation rather than investigating them. Thankfully, the report balanced this proposal by recognising the cost and the unknown risks of IVF and concluded that 'It (IVF) should not become the panacea for the treatment of all infertility causes' (32). This is wise and practical since many causes of infertility can be corrected with much simpler treatments such as drug therapy or surgery. Some couples fortuitously conceive while waiting for a clinic appointment or during their blood tests and other investigations. Many others find that the results of the various tests are normal. If a couple has been trying to conceive for a relatively short time, the normality of results may reassure them that chances of pregnancy remain optimistic, and a review appointment can be requested if couples

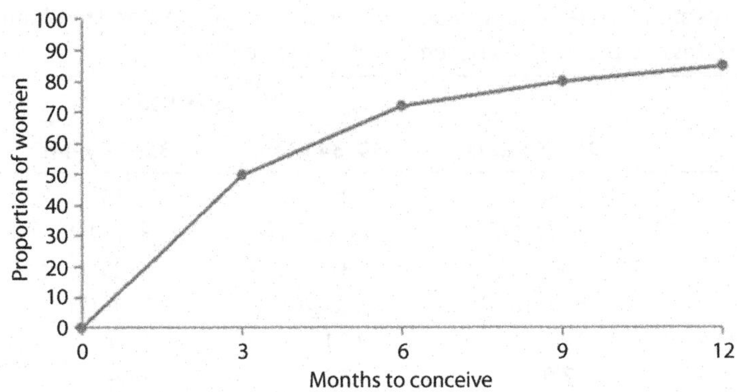

Figure 2.2 The exponential fall in monthly fecundity over time. (Adapted from data in Guttmacher AF. *Journal of the American Medical Association*. 1956;161(9):855–60.)

have not been successful after, for example, 6 months. Interestingly, the investigation to check fallopian tube patency may temporarily enhance fertility (33).

2.4 IS INFERTILITY ON THE RISE?

The number of couples seeking specialist help for infertility problems over the past 15 years has increased (22). This is partly because there is a greater awareness about the increasing range and effectiveness of infertility treatments. The increased media attention has made infertility less of a taboo, which means that friends and acquaintances are more open in talking about their own experiences of fertility problems giving rise to increased demand for interventions. There are other societal influences at work, for example many couples now delay having children. This may be for the woman to pursue her career or for a couple to achieve greater financial stability or pursue other ambitions before having a family. Twenty-five years ago, the average age for a woman in the United Kingdom to have her first child was in the early 20s; a decade ago, it was 25 years; and now it is 28.6 years for all first births, while 53% of first births are to women over 30 (23). The delay in attempting pregnancy, to an age at which the woman's natural fertility is starting to reduce, inevitably means that more couples will experience difficulty in conceiving and will seek medical advice. Natural fertility declines in women over 30, and many couples may need to consider more successful (but also more stressful and expensive) high-technology treatments to optimise their chance of pregnancy.

As female age advances, the likelihood of natural conception diminishes. Couples should be provided with precise data (about their age specific chances of conception) so as to be fully informed in their decision making. This will include information about oocyte quality as age progresses (24), but, in addition, how chromosomal abnormalities and aneuploidy contribute significantly to a fall in fertility (25). Data in Table 2.1 illustrate how this is demonstrated in a clinical IVF setting. Fewer oocytes and embryos are found in women over the ages of 35 and 40, and implantation rates and pregnancies per embryo transferred are low in women of this age group (26). Though 20 years old, these data are still valid, because although absolute pregnancy rates have improved slightly, the relative differences are the key message here.

A major factor in the aetiology of tubal disease is the damage caused by acute infectious inflammation due to Chlamydia (and maybe even more by the healing processes and heat shock proteins released as the inflammation heals). A UK screening programme, introduced in 2008, found that the prevalence of elevated Chlamydia antibody levels in women aged 20–24 years old was 21.4%, compared with 6.3% in the 16–20 year old group. The prevalence of serum antibody levels in the population increased with age, reaching a peak of 33.5% (CI 27.5–40) in the 30–34 year age group,

Table 2.1 Patients studied related to woman's age groups, oocyte yield and fertilisation and cleavage rates, implantation rates and pregnancies per embryo transferred (ET).

	Woman's age			
	25–29 years	**30–34 years**	**35–39 years**	**40–44 years**
Total oocytes	1236	2397	1855	298
Median	11 (3–55)	10 (2–31)	8 (1–26)	6 (0–26)[a]
Total embryos	660	1375	1105	191
Median	6 (0–18)	6 (0–20)	5 (0–17)	3 (0–15)
Implantation rates (%)	18.2	16.1	15.3	6.1
Mean	17.9	15.4	14.4	6.7[a]
Pregnancies per ET (%)	35	31	31	16[a]

Source: Adapted from data in Hull MG et al. *Fertility and Sterility*. 1996;65(4):783–90.

[a] All these values reach statistical significance using appropriate parametric or nonparametric methods.

and leveling off to around 30% for women aged up to 45 (27). As will be explained in ensuing chapters, raised Chlamydia antibody levels are an important and clinically useful predictor of tubal damage (28).

The media has presented a picture of falling male infertility supported by various scientific reports, which have looked at results over time and found a fall in overall sperm numbers, particularly in the developed countries of the world (20). However, there are equally as many papers insisting that there is no change in male infertility. The debate is well reviewed elsewhere (19). One explanation for the different findings may be that as examination techniques have become more exacting, sperm counts are being reported more accurately - and lower. On the other hand, it is possible that the reported trends reflect a real fall in sperm numbers, possibly because of toxins in the environment and the increasing use of steroid hormones in food production. There may even be antenatal exposures, for example, prenatal and postnatal exposure to endocrine disruptors such as the persistent environmental chemicals, Bisphenol A and p,p′-DDE. These are associated with an increased risk of male reproductive disorders and also with an increasing incidence of testicular cancers, irrespective of the histological nature of the tumour (21). Several recent reports suggest that the working environment in which people find themselves can exercise an effect on male fertility. Although they are extreme examples, nevertheless they point out the important relationship between our working environment and our fertility and the importance of remembering it as a possible detrimental effect.

2.5 SUMMARY

The prevalence of infertility appears to be rising as several factors (mostly social) contribute to what is a rising burden of infertility on individuals, health services and the economy. Encouraging women to get pregnant earlier in life is simplistic and not cognisant of life in the twenty-first century, but the impact of low income, career development and finding a suitable life partner all add to the delay in finding the right time to achieve a pregnancy.

REFERENCES

1. Hull MG, Glazener CM, Kelly NJ, Conway DI, Foster PA, Hinton RA et al. Population study of causes, treatment, and outcome of infertility. *British Medical Journal (Clinical Research Ed).* 1985;291(6510):1693–7.
2. Giri A, Edwards TL, Hartmann KE, Torstenson ES, Wellons M, Schreiner PJ et al. African genetic ancestry interacts with body mass index to modify risk for uterine fibroids. *PLoS Genetics.* 2017;13(7):e1006871.
3. Wijeyaratne CN, Balen AH, Barth JH, Belchetz PE. Clinical manifestations and insulin resistance (IR) in polycystic ovary syndrome (PCOS) among South Asians and Caucasians: Is there a difference? *Clinical Endocrinology.* 2002;57(3):343–50.
4. Huang Z, Yong EL. Ethnic differences: Is there an Asian phenotype for polycystic ovarian syndrome? *Best Practice & Research Clinical Obstetrics & Gynaecology.* 2016;37:46–55.
5. Cahill DJ, Wardle PG, Maile LA, Harlow CR, Hull MG. Pituitary-ovarian dysfunction as a cause for endometriosis-associated and unexplained infertility. *Human Reproduction (Oxford, England).* 1995;10(12):3142–46.
6. Yamamoto A, Johnstone EB, Bloom MS, Huddleston HG, Fujimoto VY. A higher prevalence of endometriosis among Asian women does not contribute to poorer IVF outcomes. *Journal of Assisted Reproduction and Genetics.* 2017;34(6):765–74.
7. Missmer SA, Hankinson SE, Spiegelman D, Barbieri RL, Marshall LM, Hunter DJ. Incidence of laparoscopically confirmed endometriosis by demographic, anthropometric, and lifestyle factors. *Am J Epidemiol.* 2004;160(8):784–96.
8. Woodworth SH, Singh M, Yussman MA, Sanfilippo JS, Cook CL, Lincoln SR. A prospective study on the association between red hair color and endometriosis in infertile patients. *Fertility and Sterility.* 1995;64(3):651–2.

9. Missmer SA, Spiegelman D, Hankinson SE, Malspeis S, Barbieri RL, Hunter DJ. Natural hair color and the incidence of endometriosis. *Fertility and Sterility.* 2006;85(4):866–70.
10. Depmann M, Faddy MJ, van der Schouw YT, Peeters PH, Broer SL, Kelsey TW et al. The relationship between variation in size of the primordial follicle pool and age at natural menopause. *Journal of Clinical Endocrinology and Metabolism.* 2015;100(6):E845–51.
11. Dolleman M, Faddy MJ, van Disseldorp J, van der Schouw YT, Messow CM, Leader B et al. The relationship between anti-Mullerian hormone in women receiving fertility assessments and age at menopause in subfertile women: Evidence from large population studies. *Journal of Clinical Endocrinology and Metabolism.* 2013;98(5):1946–53.
12. Wang JX, Davies M, Norman RJ. Body mass and probability of pregnancy during assisted reproduction treatment: Retrospective study. *BMJ (Clinical Research Ed).* 2000;321(7272):1320–1.
13. Wang JX, Davies MJ, Norman RJ. Obesity increases the risk of spontaneous abortion during infertility treatment. *Obesity Research.* 2002;10(6):551–4.
14. Dunson DB, Baird DD, Colombo B. Increased infertility with age in men and women. *Obstetrics and Gynecology.* 2004;103(1):51–6.
15. MacLeod J, Gold RZ. The male factor in fertility and infertility: V. Effect of continence on semen quality. *Fertility and Sterility.* 1952;3(4):297–315.
16. Freund M. Effect of frequency of emission on semen output and an estimate of daily sperm production in man. *Journal of Reproduction and Fertility.* 1963;6:269–85.
17. Marshburn PB, Alanis M, Matthews ML, Usadi R, Papadakis MH, Kullstam S et al. A short period of ejaculatory abstinence before intrauterine insemination is associated with higher pregnancy rates. *Fertility and Sterility.* 2010;93(1):286–8.
18. Guttmacher AF. Factors affecting normal expectancy of conception. *Journal of the American Medical Association.* 1956;161(9):855–60.
19. Kumar N, Singh AK. Trends of male factor infertility, an important cause of infertility: A review of literature. *Journal of Human Reproductive Sciences.* 2015;8(4):191–6.
20. Levine H, Jorgensen N, Martino-Andrade A, Mendiola J, Weksler-Derri D, Mindlis I et al. Temporal trends in sperm count: A systematic review and meta-regression analysis. *Human Reproduction Update.* 2017;23(6):646–59.
21. Bonde JP, Flachs EM, Rimborg S, Glazer CH, Giwercman A, Ramlau-Hansen CH et al. The epidemiologic evidence linking prenatal and postnatal exposure to endocrine disrupting chemicals with male reproductive disorders: A systematic review and meta-analysis. *Human Reproduction Update.* 2016;23(1):104–25.
22. Balasch J. Investigation of the infertile couple: Investigation of the infertile couple in the era of assisted reproductive technology: A time for reappraisal. *Human Reproduction (Oxford, England).* 2000;15(11):2251–7.
23. Office for National Statistics. Births by parents' characteristics in England and Wales: 2015 London: Office for National Statistics; 2015. Available from: https://www.ons.gov.uk/peoplepopulationandcommunity/birthsdeathsandmarriages/livebirths/bulletins/birthsbyparentscharacteristicsinenglandandwales/2015.
24. Cahill DJ, Prosser CJ, Wardle PG, Ford WC, Hull MG. Relative influence of serum follicle stimulating hormone, age and other factors on ovarian response to gonadotrophin stimulation. *British Journal of Obstetrics and Gynaecology.* 1994;101(11):999–1002.
25. Munne S, Alikani M, Tomkin G, Grifo J, Cohen J. Embryo morphology, developmental rates, and maternal age are correlated with chromosome abnormalities. *Fertility and Sterility.* 1995;64(2):382–91.
26. Hull MG, Fleming CF, Hughes AO, McDermott A. The age-related decline in female fecundity: A quantitative controlled study of implanting capacity and survival of individual embryos after *in vitro* fertilization. *Fertility and Sterility.* 1996;65(4):783–90.

27. Woodhall SC, Wills GS, Horner PJ, Craig R, Mindell JS, Murphy G et al. *Chlamydia trachomatis* Pgp3 antibody population seroprevalence before and during an era of widespread opportunistic Chlamydia screening in England (1994–2012). *PLOS ONE*. 2017;12(1):e0152810.
28. Akande VA, Hunt LP, Cahill DJ, Caul EO, Ford WC, Jenkins JM. Tubal damage in infertile women: Prediction using Chlamydia serology. *Human Reproduction (Oxford, England)*. 2003;18(9):1841–7.
29. National Collaborating Centre for Women's and Children's Health (2013). *National Institute for Health and Clinical Excellence: Guidance. Fertility: Assessment and Treatment for People with Fertility Problems*. London, Royal College of Obstetricians & Gynaecologists.
30. Snick HK, Snick TS, Evers JL, Collins JA. The spontaneous pregnancy prognosis in untreated subfertile couples: The Walcheren primary care study. *Human reproduction (Oxford, England)*. 1997;12(7):1582-8.
31. Mascarenhas MN, Flaxman SR, Boerma T, Vanderpoel S, Stevens GA. National, regional, and global trends in infertility prevalence since 1990: A systematic analysis of 277 health surveys. *PLoS medicine*. 2012;9(12):e1001356.
32. ESHRE Capri Workshop Group. Economic aspects of infertility care: A challenge for researchers and clinicians. *Hum Reprod (Oxford, England)* 30(10): 2243–8.
33. Mohiyiddeen L, Hardiman A, Fitzgerald C, Hughes E, Mol BJ, Johnson N, Watson A. Tubal flushing for subfertility. *Cochrane Database of Systematic Reviews* 2015, Issue 5. Art. No.: CD003718. DOI: 10.1002/14651858.CD003718.pub4.

Investigations in infertility 3

OVERVIEW

This chapter covers the basic and advanced investigations to use in infertile couples, the rationale behind those tests, the treatments that are presently possible for different diagnostic categories and the need for the sensitive management of couples because of the personal nature of this condition. It includes a brief guide to conducting a first interview, the tests to undertake and their interpretation and some case studies to illustrate how these might be put into practice.

3.1 CONSULTATIONS

Infertility can be extremely distressing for couples. As a clinician working with these patients, you will require a certain skill set seasoned with compassion and understanding to provide as efficient and sympathetic a service as possible. Couples have a need for information, emotional support and counselling, as well as investigations and treatments. Sharing problems with other couples in self-help groups can be valuable. In my clinical service, the nurses play a critical role in managing and supporting patients. They are not just chaperones or handmaidens; they manage many of the treatment processes independently, liaising as required, but taking day-to-day responsibility for patient care (1). I find it difficult to think, having worked in that environment for 20 years, how else one could run an efficient service, and I would encourage readers to consider that approach as being an ideal (if not gold standard) manner to run a clinical service.

3.1.1 There are several goals when seeing a couple in clinic

- *To optimise* the infertile couple's health in preparation for becoming pregnant, both to improve chances of conceiving and to maximise the health of a baby during pregnancy. This includes advice about diet and smoking and immunisation against infections (such as rubella, or German measles – though population immunity for rubella is high, and children vaccinated with the MMR should be immune). There are some fertility clinics which require both partners to stop smoking for 6 months to be eligible for investigations or treatment and before commencing any workup.
- *To diagnose* the possible cause or causes of infertility by all the necessary investigations, starting with the simplest.
- *To counsel and advise couples* by providing all the information they need to make their own decisions about possible treatments, and, if necessary, to help them consider their fertility needs in balance with other important parts of their lives.
- *To treat couples* by whatever methods best suit their requirements and personal wishes in choosing between alternative treatments.
- *To protect* any baby born as a result of treatment from avoidable risks, such as infections and genetic defects and from unhealthy parental lifestyles or behaviour.

3.1.2 Suggested plan for initial consultation, investigation and management

The couple should be reviewed when the results are available – within a publicly funded health service this may be 6–8 weeks later or more. It should be relatively evident from the investigation results which diagnostic category/ies a couple falls into. Management will then be along the lines of the options suggested by the history, if helpful, and the investigations. The approach to management will be tempered by factors such as age which might, if advanced, be a reason to accelerate through treatment options to the most effective options (normally assisted conception).

3.1.3 Service delivery constraints in national public health services such as the United Kingdom's National Health Service

Provision of services in National Health Service (NHS) infertility clinics varies according to local clinic commissioning group criteria as to the eligibility of patients to obtain infertility services, though the guidelines published by the National Institute for Health and Care Excellence (NICE) will be the foundation for all clinic commissioning group decision making (2). As a generalisation, provision in Wales and the north of England is broader than in the south. Criteria used to select patients for assessment are many and varied, but can include age, usually being less than 40 years for the woman; BMI, usually less than 30 kg/m^2 for the woman; previous sterilisation for either partner; nonsmoking for both partners (established by the use of a carbon monoxide monitor at first visit); duration of relationship greater than 2 years – obtained on history; no recreational drug use – obtained on history; any previous children from the relationship limits access to the service. All of these are designed to restrict access, and some have a scientific basis (age, BMI, smoking, for example).

3.2 THE FIRST APPOINTMENT

3.2.1 History taking

At first consultation, both partners should be asked to attend as a couple. This is a couple problem. Together, they are unable to achieve a pregnancy. In the rest of this account, the couple will be referred to as male and female, as this is the predominant pattern of referrals to clinics. Male problems and female problems are equally applicable to couples of the same or opposite sex. Single individuals may recognise that they have problems with their fertility and seek advice and treatment. Such individuals are unlikely to be funded by the NHS, but they may well present to private medicine.

Medical histories should be taken in detail, looking at pubertal development, past fertility, past abdominal surgical history and family history of PCOS, endometriosis and infertility. The man's history should rule out past scrotal injury, mumps and vasectomy. In our initial correspondence with couples, we ask women to bring a list of the starting dates of their last six menstrual periods, if they can produce them. This is an important piece of information because regular cycles suggest regular ovulation and provide guidance about when to do tests about ovulation. Both the man and the woman will have a physical examination, partly focussed by the history, but including an ultrasound scan of the pelvis for the woman and scrotal examination for the man. The first visit is the opportune time to do these, although it may prove embarrassing to one or both. Sensitive, assertive explanation as to the importance of these examinations is vital.

A plan of action for this initial consultation, investigation and management is described here:

- See the couple who are trying to conceive as a couple.
- Take a history from both.
- Focus on age, smoking, alcohol intake, prescribed and social/recreational drug use, present and past weight, previous pregnancy, past contraceptive use and abdominal or genital surgery (every one of these has a direct effect on the likelihood of any further pregnancies).

3.2.2 Physical examination, including ultrasound scans

The couple should be examined separately, with independent chaperones. There is an advantage to seeing the couple individually in this way, to allow them to divulge any very sensitive information, and this should be facilitated.

Examination of the man includes his external genitalia – the scrotum, the testicles, the epididymis and the vas deferens leading from the testis up to external opening of the inguinal canal, excluding any evidence of bowel herniation which might increase the scrotal temperature. The aim is to estimate testicular size and volume, to confirm the presence of a vas deferens in its entirety and

Table 3.1 Investigations for infertility in women and men.

Test	Reason to undertake this test	Conditions required for the test	Normal values and units (in our laboratories)
Hepatitis B, C screening; HIV and HTLV screening, syphilis	Avoiding infection of the fetus in pregnancy Most laboratories would want to know these details before handling and certainly before freezing any gametes	Any time or day	No evidence of infection
Rubella immunity	Infection in pregnancy causes major fetal anomalies		Evidence of immunity
Full blood count	Rule out anaemia prior to pregnancy to prevent adverse effect on mother and fetus	Any time or day	A haemoglobin level >12 g/dL or 120 g/L and other normal indices
FSH (in women)	Examines the ability of the ovary to produce follicles, assesses the reserve supply of eggs and, with LH levels, predicts polycystic ovary syndrome	Best in the first week of the woman's cycle	1–9 IU/L
LH (in women)	With FSH levels, predicts polycystic ovary syndrome	As for FSH	1–9 IU/L
AMH	An accurate measure of ovarian reserve	Any time or day	Levels below 4–5 pmol/L are associated with poor ovarian response
Antral follicle count	An accurate measure of ovarian reserve	Best in the first week of woman's cycle	At least six follicles 2–6 mm diameter
Progesterone	Determines if ovulation took place	Best done 5–10 days before a period is due	>30 nmol/L (conversion factor 3.1446 to ng/mL)
TSH	Detects an underfunctioning thyroid gland, which can affect fertility	Any time or day	0.3–3.0 IU/L
Prolactin	A hormone which, if overproduced, suppresses ovulation and menstruation	Any time or day	100–700 mU/L

(Continued)

Table 3.1 (Continued) Investigations for infertility in women and men.

Test	Reason to undertake this test	Conditions required for the test	Normal values and units (in our laboratories)
Semen analysis	Checks for the presence and quantity of sperm	Examined within an hour of production	Count >20 million sperm/mL of sample, Normality >4% Motility >50%
Postcoital test	Checks the function of sperm and the quality of cervical mucus	Examined 8–12 hours after intercourse at midcycle, with cervical mucus stretching to at least 10 cm	Normal: >3 motile sperm Impaired: 1–3 sperm Negative: 0 sperm pH 6.5–8.5 All per high-power microscope field (400× magnification)
Chlamydia antibody titre	Used to determine the likelihood and severity of tubal damage and whether HSG or laparoscopy is indicated to review tubal patency (see Figure 3.1)	Any time or day	Antibody titres ≤1:256 indicate low risk of tubal damage, ≥1:512 indicate high risk of tubal damage

Source: Wardle PG, Cahill DJ. *Understanding Infertility*. Family Doctor Publications, Poole, United Kingdom; 2005, with permission.

Abbreviations: FSH, follicle-stimulating hormone; IU/L, international units per litre; LH, luteinising hormone; mU/L, milliunits per litre; AMH, anti-Mullerian hormone; TSH, thyroid-stimulating hormone; HSG, Hysterosalpingogram.

to exclude any varicoceles in and around the epididymis. The evidence that bowel herniation and varicoceles impair male fertility is very slim, but it is worth noting their presence.

The most effective examination of the woman is by a transvaginal examination of uterus and ovaries. A transvaginal scan as part of a woman's first visit is perhaps daunting for her, but the quality of information is far superior to either bimanual examination or transabdominal scanning. Close and detailed inspection can be made of the uterus and endometrium and the endometrial cavity, as well as estimation of the antral follicle count (see Section 3.5.1) and appearance of the ovaries. Bimanual examination is unlikely to reveal any further information that would assist the diagnosis [the impact of uterine retroversion on fertility is largely refuted in the absence of endometriosis (3)].

Investigations should include the range of blood and other tests provided in Table 3.1. The results of any previous investigations will be reviewed. If considered reliable, they will not be repeated.

3.3 INITIAL INVESTIGATIONS: FOR THE WOMAN

3.3.1 Investigations

- Blood samples for various antibodies, hormone measurements and iron stores levels.
    i. Antibodies for Hepatitis B and C, human immunodeficiency virus (HIV) 1 and 2, syphilis and rubella (German measles)
    ii. Full blood count and ferritin levels
    iii. Chlamydia antibody blood test
    iv. Thyroid-stimulating hormone and prolactin (if cycles are more than 42 days apart)

- A blood sample during the first 5 days (Days 1–5) of the menstrual cycle for FSH and LH measurements. This test should be repeated in another one or two cycles if there are abnormal results, to confirm the abnormality, because the hormone levels can vary significantly from cycle to cycle. You may consider offering a blood test for anti-Mullerian hormone (AMH) (see Section 3.3.2). This, like FSH levels, is a test of ovarian reserve, but its value is still debated.
- A blood sample towards the end of the menstrual cycle (7–10 days before the next menstrual period is expected), examining for progesterone hormone, to check on the quality of ovulation. In the same way as for the previous tests, this may need to be repeated in another one or two cycles if there are abnormal results. If women have irregular cycles, the best course is to take a blood test at Day 21 and repeat weekly until a period comes. It is very likely that the result will not show evidence of ovulation.
- Collection of a cervical mucus sample just before ovulation is due and the day after coitus, when the cervical mucus is optimal, will allow the clinician to do a postcoital test (PCT). Timing of the test in the cycle is important and it often needs to be repeated. If timing proves difficult and if the couple have to travel some way to the clinic (thereby adding to the inconvenience), it is sensible to prescribe some oral oestrogen tablets for the woman to control the timing of the mucus to facilitate this testing (50–100 μg oral ethinyl oestradiol for 5–7 days will usually produce good-quality cervical mucus).

3.3.2 Rationale for testing: For the woman

Viral screening for hepatitis, HIV and syphilis should be undertaken to ensure the safety of any baby. All three conditions are transmitted across the placenta. Furthermore, given that any couple presenting to a clinic may need to have embryos or eggs and sperm stored in freeze storage, it follows that all should be subject to screening prior to entering into the process to avoid cross-contamination with the stored products. It allows any couple identified as having viral antibodies to be investigated by virologists/hepatologists and treated if possible.

Checking the woman for rubella antibodies means the identification of those susceptible to German measles for whatever reason (never having the infection, never getting the vaccination as a child, not developing immunity after vaccination). This allows for vaccination and appropriate management before entering on any treatment. Advice if immunity doesn't develop after a second vaccination would generally be to proceed with the infertility treatment, avoiding infection if possible.

Checking haemoglobin levels and iron stores is one measure of the woman's general nutritional status and readiness to carry a pregnancy. There is some weak evidence (never tested in a randomised controlled trial) that conception is limited in women with depleted iron stores (4).

A positive finding of Chlamydia antibody is an important prognostic factor. Research in Bristol, studying 1000 women, found a direct correlation between Chlamydia antibody titres and tubal damage, with an increased likelihood of damage the higher the titre readings (5). The Chlamydia risk table (Figure 3.1) illustrates the relationship between tubal damage and Chlamydia titre levels. This information is used to guide choice of further examination, either laparoscopy with dye hydrotubation or hysterosalpingography. Women whose titres are low or negative (<1:64–1:128) have a hysterosalpingogram. Women whose titres are higher (≥1:512) and therefore considered at greater risk of severe tubal damage proceed directly to having a laparoscopy and dye hydrotubation. Women whose titres are in between will have one or other investigation, depending on the presence or absence of a history of pelvic infection and their partner's sperm quality. The Bristol findings echo Westrom's research many years ago showing that repeated episodes of pelvic inflammatory disease (PID), from whatever cause, increased the risk of tubal occlusion, and the risk of damage increased almost exponentially as the frequency of episodes of PID increased (6).

Thyroid stimulating hormone and prolactin. TSH and prolactin should not be measured if a woman has normal periods, as it is very unlikely that the woman will have any significant ovarian dysfunction, and the tests will be most likely normal (7). Raised prolactin levels are likely to give

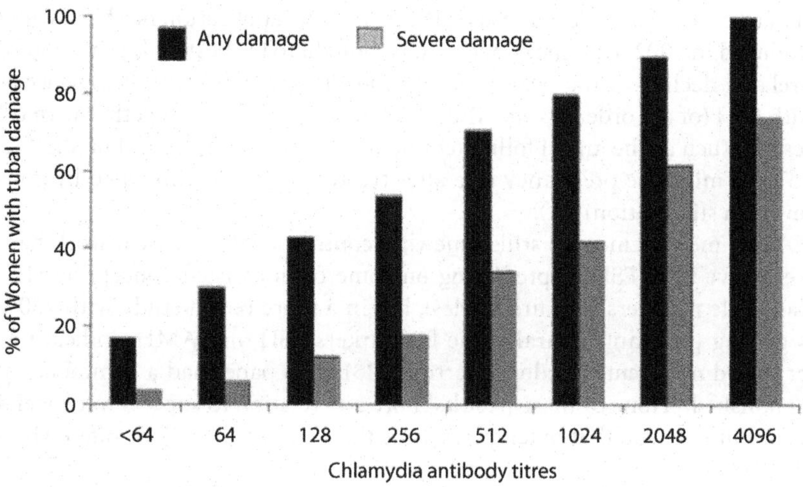

Figure 3.1 Frequency (%) of any tubal damage or severe tubal damage according to Chlamydia antibody titres. (From Akande VA et al. *Human Reproduction*. 2003;18(9):1841–7, with permission.)

rise to absence of periods and are the cause of amenorrhea in up to 15% of women. Irregular periods may occur at lower abnormal levels (1%–2% of women with irregular cycles). In women with hypothyroidism, up to 60% will have irregular cycles and anovulation.

FSH and LH on Days 1–5 of the menstrual cycle. Measuring FSH and LH levels in Days 1–5 of the follicular phase of the cycle is a fundamental step to understanding the endocrinology of ovarian function in any woman. FSH provides information on the amount of effort required by the anterior pituitary to drive ovulatory function, recruiting and then growing follicles to maturity. The term used to describe the amount of function left in the ovary for follicular development is ovarian reserve. LH levels generally (in normal situations) should be relatively closely aligned with the FSH level.

The normal FSH range given by most UK laboratories is 1–10 IU/L. If a woman's results fall within that range, we would expect them to have normal ovarian reserve and function, with normal and regular periods, and function normally under gonadotrophin stimulation (8). In women with irregular or absent periods, FSH levels are more likely to be abnormal, either unmeasurable/hovering at the bottom of the normal range or at the upper end or beyond. Very low levels of FSH tend to be associated with an inadequate stimulation of the anterior pituitary by GnRH (secreted by the pituitary), resulting in low levels of FSH, and then low levels of oestradiol. This condition is called hypogonadotrophic hypogonadism – Chapter 5 will address treatment options. FSH levels can also be at the high end of the normal range or higher. In this case, the expectation is that pituitary drive to the ovaries is high, but the ovaries are not responding. Most frequently, this represents a low ovarian reserve (or even ovarian exhaustion), with few follicles with ovulatory potential left in the ovary. This is called incipient ovarian failure, and it is likely to lead to premature menopause in the near future (how quickly is not possible to predict with FSH alone – see AMH and antral follicle count) (see below and Section 3.5.1).

LH has the same normal range as FSH in most UK laboratories (1–10 IU/L). As with FSH, very low levels of LH are associated with hypogonadotrophic hypogonadism. Conversely, LH levels tend to lag behind FSH as FSH levels rise in incipient ovarian failure, when LH levels can remain in the normal range for months. When LH levels are elevated alongside normal FSH levels, consider polycystic ovarian syndrome, even though this biochemical abnormality is not part of the European Society for Human Reproduction and Embryology/American Society of Reproductive Medicine (ESHRE/ASRM) classification (9). An explanation for the background to this condition is provided in Chapter 5.

AMH as a marker for ovarian reserve is relatively new. Its application has been widening since it was first introduced in 2002. In some quarters, it is thought to be 'best endocrine marker for assessing the age-related decline of the ovarian pool in healthy women' (10). It is relatively expensive compared with FSH (of the order of – in GBP – £50–£100 vs £5). Used together with other markers of ovarian reserve such as the antral follicle count (see Section 3.5.1), useful predictive models can be derived to determine the pregnancy rate after treatment cycles stimulated by gonadotrophins (controlled ovarian stimulation) (11).

However, AMH measurement is still somewhat controversial. Our own work has shown it to be no more effective than FSH in predicting outcome from assisted conception (12). This study included small cycle numbers (107 first cycles), but in a more recent study with 750 women aged 30–44 years getting pregnant naturally, the biomarkers FSH and AMH, indicating diminished ovarian reserve, did not identify reduced fertility (13). This paper had a commentary added to it when it was published: 'None of these [results] bore any relation to success in conceiving over the following years. But I bet you that in ten years' time the[se tests] will still be offered by most private fertility clinics' (14).

Progesterone is a hormone produced by the corpus luteum as a result of ovulation and change in function of the follicle after the release of LH prior to ovulation. It is a recognised marker of ovulation and more reliable than other ovulation markers such as urinary LH, changes in cervical mucus and so on. Tests for progesterone should done during a specific time window in the menstrual cycle, centred on Day 21 of the cycle. We ask women to ring our clinic when their next period starts in order to schedule the test for 7 days before the expected start of the following menstrual period. That accuracy of timing is often difficult to achieve.

UK laboratories tend to describe the lower level of normal for progesterone as 30 nmol/L, but I would accept any level above 25 nmol/L as being acceptable for biochemical evidence of ovulation. How can progesterone be measured when periods are irregular and unpredictable? One suggestion is to start measuring progesterone 21 days from the start of the current period and then measure it weekly until a period comes. If, as is likely, these irregular extended cycles are not ovulatory, this approach will confirm that.

The use of the postcoital test is discussed in detail, regarding its reliability and the practical undertaking of the test, in Chapter 4 (Section 4.7). Many clinics now dismiss this as unpractical or not meaningful.

3.4 INITIAL INVESTIGATIONS: FOR THE MAN

3.4.1 Investigations

- *Seminal fluid collection and microscopic analysis* assess sperm numbers, motility and morphology. Lower limits for this test are outlined in Chapter 4 (Section 4.1). A sperm sample should be obtained even with a history of vasectomy. The semen sample needs to be collected in the hospital/clinic, or at home, by masturbation into a special plastic container. If masturbation is not practiced by the man, or if production into a sample pot is difficult, most clinics can provide nonspermicidal condoms to use during intercourse and to collect the sample that way. It must not be collected into a normal/regular condom, as the spermicide in the condom will impair the motility of the sperm and distort the results. The semen should be examined ideally within 1 hour of production to check sperm motility. This requires advance planning on the part of the couple, with intercourse or masturbation occurring at a time when the sample can be delivered to the laboratory at a time when it can be examined promptly. Sometimes a second sample may be needed if the first one was of poor quality, sometimes due to it being only a partial sample.
- *Blood sample* for viral antibody tests and hormone measurements.
    i. Hepatitis B and C antibodies, HIV 1 and 2, syphilis
    ii. FSH, LH and testosterone if there is any suggestion of sperm disorder

3.4.2 Rationale for these tests: For the man

Seminal fluid analysis' is a basic and simple test to begin with. It gives a sperm count, the number of sperm in the sample and some information about appearance of the sperm and motility. It does *not* tell us about sperm function, which is whether or not the sperm could penetrate an egg. Tests of sperm function, in which interactions between sperm and biological materials (such as cervical mucus) have been studied, have been found to be good predictors of the likelihood of spontaneous conception, much more so than sperm counts (15). Complete absence of sperm is rare, and, together with very severe low sperm numbers, should be investigated (see Section 4.2 in Chapter 4).

Sperm counts are usually the first line of investigation; they are less time consuming and demanding of laboratory time. Sperm function tests (more detail in Section 4.1.1) *are* time consuming and demanding of laboratory time; therefore, economics and pragmatism tend to reserve tests of function for clinical scenarios in which that might be important and relevant.

Viral screening for hepatitis, HIV and syphilis are undertaken for the same reasons as for women (above).

3.5 FURTHER CONSULTATION AND INVESTIGATIONS

The couple should be seen for a second time once all test results are available. This may take up to 8–12 weeks because several investigations are menstrual cycle dependent, even in the most efficient clinics. The review appointment provides an opportunity to discuss the results and to decide on any further investigations to reach a final diagnosis.

It should be relatively evident from the investigation results which diagnostic category/ies a couple falls into. Management will then be along the lines of the options suggested by the history, if helpful, and the investigations. The approach to management will be tempered by factors such as age, which might, if advanced, be a reason to accelerate through treatment options to the most effective options (normally assisted conception).

3.5.1 Further investigations may include

- More hormone tests – progesterone assays often need to be repeated, as the first one may be poorly timed. Because of the serious implications of a raised FSH (>10 IU/L), any raised level should be retested in a following cycle, as levels can vary somewhat from cycle to cycle. Unresolved debate exists as to which value should be used in case of such variation, but many would consider the highest value to be the one to use in decision making.
- Ultrasound scanning (if not done at the first visit) of the ovaries and uterus. If the initial ultrasound scan was suggestive of uterine fibroids, then further investigation by MRI scanning of the pelvis will determine the number and extent of the fibroids. Antral follicle count (AFC), a relatively new method to assess the numbers of follicles between 2–6 mm in the first week after menstruation (16), has been compared to AMH to determine which is a better indicator of ovarian reserve – AMH would appear to be marginally better than AFC (11).
- Ultrasound scanning and specialised radiological assessment of the vas deferens are very rarely indicated, and should be requested only after taking advice from a urologist.
- Laparoscopy and dye hydrotubation directly examines the woman's tubes, ovaries, uterus and related structures. The procedure involves passing a narrow telescope into the abdomen through a 12-mm subumbilical incision under general anaesthesia. This is usually done as a day case. I would generally combine this with a hysteroscopy, allowing direct uterine access, looking at the interior anatomy of the uterine cavity. Abnormality of the cavity, polyps and submucosal fibroids may be detected using hysteroscopy (see Chapter 6). Hysteroscopy involves passing an even narrower telescope (5 mm or less) along the cervical canal and can be conveniently combined with the laparoscopy.
- Hysterosalpingogram (HSG) is a radiological examination of the uterus and fallopian tubes. This needs to be timed soon after the menstrual period finishes, and instructions should be provided to the patient about how to arrange it (see Figure 6.5).

- Contrast ultrasonography. This is an ultrasound-based assessment, with the advantage that it can be done at any time of the cycle, unlike a hysterosalpingogram; it doesn't expose the woman to ionising radiation and is better tolerated than hysterosalpingography. It requires highly trained sonographers to obtain good-quality images. In my experience, it provides images which do not appear as crisp or clear compared to hysterosalpingography. However, with good sonography skills, it provides good sensitivity and specificity in the investigation of tubal function (17).
- More specialised sperm function tests – these will be indicated if sperm laboratory reports show considerably abnormal tests in the total number, motility or proportion of normal sperm in the sample. If any of these three indices is markedly suboptimal, proceed to further tests of sperm function. In my opinion, one of the best tests for this, in terms of being straightforward and consistent across laboratories, is a test called *sperm recovery and survival.* It involves production of another sperm sample, repeating the test of number, motility and normality of the sample. Then the sperm sample is put through a 'recovery' process which is designed to remove debris and white blood cells and leave only a pellet of highly concentrated sperm. After the recovery process, the proportion of motile sperm is assessed, after which the sperm sample is left in an incubator overnight. Examination the next day determines how many motile sperm *survive*, termed the *total motile sperm* (TMS) count. Although this test is reproducible and employed across laboratories in the United Kingdom, the results must be interpreted and understood according to local lab conditions and practice and to the local outcomes from intrauterine insemination (IUI) and *in vitro* fertilisation (IVF).
- If the man is azoospermic or oligozoospermic, repeat the simple semen analysis. Arrange FSH and LH assays and testosterone. These should be done between 8 a.m. and midday, as this is the period of least biological variability. Arrange chromosomal analysis, looking particularly at the Y chromosome, looking for deletions and ruling out the XYY configuration (Klinefelter's syndrome), as well markers for cystic fibrosis.
- Repeat external examination of the scrotum, paying attention to testicular volume, epidydimal masses, varicocoeles and the normality of the palpable vas deferens.

3.6 CONSULTATION TO DECIDE DIAGNOSIS AND DISCUSS TREATMENT

This is arranged when all the required investigations have been done to reach a diagnosis. Ideally this should happen no longer than 3–6 months after initial consultation. The couple needs an explanation of the test results, advice about their options in the light of the tests results and time to consider how they want to proceed. Not all couples need treatment; some may be helped by advice alone.

As outlined in Chapter 2, there are four main diagnostic categories.

1. Tubal damage
2. Ovulatory dysfunction
3. Sperm dysfunction
4. Unexplained and endometriosis

The relative prevalence of each one (from several studies) is listed in Table 3.2. Unexplained infertility and endometriosis are given as a diagnostic category. Endometriosis is often disputed by some as a cause of infertility. Published data from Bristol and elsewhere have consistently shown a negative effect of endometriosis on fertility in conception (18) and assisted conception (19–21). To make the diagnosis of unexplained infertility (examined fully in Chapter 7) one must establish that: the menstrual cycle was normal (21–42 days in length), seminal analysis of the man yielded normal results, coital activity with full penetration and vaginal ejaculation occurred at least twice a week and tubal patency was normal [assessed by laparoscopy or hysteroscopy (22)].

There are further uncommon causes of infertility outlined elsewhere (22). Some are interesting, even intriguing, for example, failure of postcoital sperm penetration of the mucus. Before assisted conception techniques were well developed (IUI, for instance), it was important to identify the cause and attempt specific corrective measures (a sodium bicarbonate douche in the vagina before intercourse in the case of acidic mucus.) Nowadays, with the wider availability of assisted conception

Table 3.2 Analysis of the diagnoses arrived at in studies.

	UK (n = 708) (%)	Canada (n = 2198) (%)	Netherlands (n = 726) (%)
All male factor	26	27	30
Azoospermia	6	7	5
Ovulatory	21	27	26
Tubal	21	27	26
Endometriosis	6	14	3
Unexplained	28	6	30

Source: Collins JA et al. *Fertility and Sterility*. 1995;64(1):22–8; Snick HK et al. *Human Reproduction (Oxford, England)*. 1997;12(7):1582–8; Hull MG et al. *British Medical Journal (Clinical Research Ed)*. 1985;291(6510):1693–7.

Note: In three different countries, the largest proportion in two of these three is unexplained infertility – meaning all the appropriate investigations have been done and no obvious cause has been identified.

interventions, the usual practice is to identify that the sperm, although healthy in themselves, are not achieving fertilisation of (already established) healthy oocytes and proceed to assisted conception without identifying the precise cause of the problem. Specialists tend to finish the investigatory phase sooner than was the practice in previous decides, and they move to on to therapeutic intervention more quickly. This problem would now be subsumed in the 'male category' when listing the cause of the infertility problem.

3.7 TREATMENT OPTIONS IN OUTLINE (MORE DETAILS IN CHAPTERS 4–7)

3.7.1 General advice

All couples, regardless of diagnosis, should be advised how to maximise their chance of achieving a pregnancy. This should include advice on diet, smoking reduction or cessation (25), weight reduction to a BMI between 20 and 25 and taking folic acid regularly. As much effort and importance should be put into explaining the benefit of these measures to the couple as one would give when talking them through medical or surgical interventions. The public can be deaf to exhortations to adopt a healthy lifestyle. The clinician must convince the couple of the direct value to themselves and their fertility to do so.

3.7.2 No treatment indicated

Some couples' investigations will be essentially normal. The diagnosis will be given as 'unexplained infertility'. The couple should be encouraged about the normality of the results and their healthy status and assured that they have a good chance of their conceiving naturally. Once the couple have been trying for 3 years, the couple should be offered active intervention such as IUI or IVF.

3.7.2.1 For the woman

- All forms of treatment for hormone disorders and ovulation failure.
- All forms of surgery including microsurgery and keyhole laparoscopic surgery, to repair blocked tubes and other pelvic conditions (such as fibroids). The relatively rare finding of proximal tubal blockage can be diagnosed and treated at the same time by selective salpingography, a specialised form of the hysterosalpingogram examination. It involves direct catheterisation of one or the other fallopian tube, passage of dye and then, in the presence of a proximal tubal blockage, passing a soft flexible 'wire' under x-ray guidance (fluoroscopy) to unblock the tube at its inner end (Figure 3.2).

3.7.2.2 For the man

- There is no convincing evidence that medication options such as nutritional supplements, aromatase inhibitors or other endocrine-focussed therapies have any beneficial effects on the likelihood of natural conception.

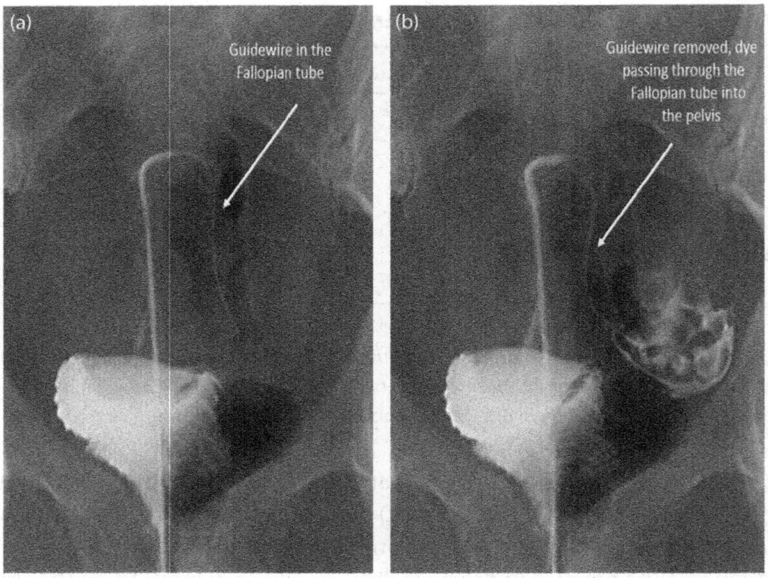

Figure 3.2 Selective salpingography of the left fallopian tube, with the guidewire being passed into the tube (a) and then, with the guidewire removed (b), radio-opaque dye is injected, running along the tube and into the pelvis, demonstrating patency.

- If azoospermia or sperm dysfunction is present, one treatment option is to use artificial insemination with normal donor sperm. In the United Kingdom, this treatment is only available through clinics licenced by the Human Fertilisation and Embryology Authority. Access to donor sperm in other jurisdictions may be less regulated.
- In the rare event of a blocked or damaged vas deferens, surgery may be effective but should be managed by referral to a urologist. Surgical procedures to collect sperm from the epididymis or testis are generally done in IVF clinics (see Section 4.6.2 in Chapter 4, for more details).
- A minority of men are anejaculatory because they have had a spinal cord injury with disruption of the nerve supply supporting erection and ejaculation, easily managed by electro-ejaculation or needle aspiration of the epididymis (see Section 4.6.2 for more details).

3.7.2.3 For the couple ('assisted conception')

In vitro fertilisation treatment is effective for several causes of infertility. When a couple have been trying to conceive for several years, no matter what the diagnosis is, IVF remains the default position as the treatment of choice in terms of being effective. This is particularly the case with older women, where the impact of age is highly significant in contributing to failing to conceive. In the case of sperm problems, a special technique in IVF is injecting a single sperm – selected for shape, mobility and similar relatively subjective criteria – into the egg to achieve fertilisation (intracytoplasmic sperm injection, or ICSI). Access to ICSI is often limited by cost – availability of funded treatment across Europe varies considerably. Not surprisingly, access to assisted conception for couples is often directly related to their ability to pay (23).

3.8 COUNSELLING

3.8.1 Impact of the diagnosis

If the cause of infertility has been diagnosed, the clinician should be aware of and look for any psychological impact in one or other partner. A low sperm count which offers little hope of cure can seriously affect a man's self-image. He may feel less of a man now; he may feel impotent, or that lovemaking is pointless. It is important to explain, when giving the test results, and to repeat it again in the session, that male infertility has little to do with hormones and virility. Chapter 4 will expand on this. Fallopian tube damage and high Chlamydia titres can cause questioning and recriminations between the partners about previous sexual behaviour. The source of that infection leads to questioning within the couple if they have been in a stable relationship for many years. It may create doubt in their minds as to the faithfulness of the other, and emphatic explanation must be given by the clinician that the infection could have started many, many years ago and predate their relationship.

Infertility can cause great sadness and anxiety for the individuals involved. It can lead to severe strains on a couple's relationship, especially if one partner thinks that the other isn't committed to the endeavour to have a child or if one holds the other responsible for the problem. Cultural or family attitudes to children add to this. For some, the very personal intrusiveness of some of the questions and investigations imposed on couples leads to resentment or withdrawal. Months of appointments, time off work, transport and parking problems, blood tests, timing of blood tests, scans, painful internal investigations and producing sperm samples to order, waiting and hoping and, in the private setting, the financial burden – everything involved in the investigation and treatment process adds to the couple's stress. All clinical staff should be trained in awareness of these issues and in how to provide sensitivity, general support, explanation and encouragement to each partner. Special counselling should be available to help couples cope with the particular sadness and stress of longing for a baby, which can be such a dominating factor in the lives of couples with fertility problems. This is of such importance that access to counselling is something that all clinics in the United Kingdom licenced to undertake assisted conception treatments have to provide. Reassuringly, the evidence for any negative effect of general, or infertility-related, distress on the chances of conception from treatment is not convincing (24).

3.8.2 Impact of treatment options

Moving on to the treatment phase, self-administration of the drugs – gonadotrophin injections – can be stressful as the patient agonises about the timing and correct technique. Arranging the necessary blood tests and scans can be problematic for the patient – and clinical staff – if the patient wishes to conceal the treatment stage from their work colleagues. Often the male partner struggles, because of work, to be available on the day and at the correct time of the day to produce the sperm sample for IUI or IVF. If the follicles don't grow or if the embryo doesn't implant, the woman may fret about what she is doing wrong (and of course, probably everything she is doing is fine).

The use of donor gametes may involve a considerable shift in the couple's personal expectations for their biological link to their child. A couple would be expected to need some time to come to terms with this and to be able to discuss this with an independent counsellor (beyond their clinician).

3.9 CASE STUDY WITH A COMMENTARY

Roger and Poppy have lived together for 4 years. Poppy was recently promoted at work and moved office to be nearer home. They have decided that they want to try for a family. Poppy is 31, of normal weight, and her BMI is 26. She doesn't smoke and rarely drinks, a glass of white wine when she does. Her periods are irregular and have been for 2 years. They have used condoms and a diaphragm for contraception, as Poppy doesn't like the idea of using chemicals to affect her body. After 9 months of unprotected intercourse, she is getting anxious because she is not pregnant. She recently used ovulation kits to determine when she is ovulating and when to have intercourse. The kits seem to suggest that she is not ovulating. She makes an appointment to see the GP, who points out that she has been trying for only 9 months and she should wait at least a year before doing anything more. Poppy persuades the GP to do some blood tests – and these are done over the next few weeks on different days. A few weeks later, she returns with Roger to get the blood results and to try to arrange a sperm test for him.

The GP takes time explaining the results. She tells the couple that one of the blood tests is very abnormal and indicates that Poppy is almost menopausal. Poppy struggles to understand this and cannot believe it. The GP says that in the light of these results, she is going to arrange an urgent appointment for Poppy and Roger to see a specialist. The appointment is a month later. The doctor they see confirms the serious nature of the blood result and arranges a scan of Poppy's ovaries the same day, and a blood test. The results of the blood test and the scan are discussed at the next visit, which is another month later. The specialist gently imparts the news that Poppy is about to start her menopause – all the blood tests agree on this and the ovary scan report says there were very few follicles in her ovaries. The specialist offers Poppy a referral to a local private IVF clinic to discuss IVF treatment and tells her she should do this right away.

Commentary: Poppy's results show a raised FSH level, at 22 IU/L. The further blood test was an anti-Mullerian hormone test. The result was 6.7 pmol/L, which is low and in line with the raised FSH level. The ultrasound scan of ovaries looked at the antral follicle count and found very low numbers of follicles. The history of irregular cycles for 2 years, with these test results, indicates the diagnosis of impending early menopause. The low antral follicle count means that the chances of her responding to ovulation stimulation drug injections are poor.

This is a very difficult situation for anyone to cope with. Poppy and Roger need time to come to terms with their threatened hopes and expectations. At the same time, they must make decisions about attending an IVF clinic, with the financial burden this will bring. An alternative way to manage this is for the specialist to see them again after a short interval and provide realistic information about what is likely to happen and about her chances of getting pregnant. If she goes ahead with IVF treatment (using her own eggs), her FSH level indicates that she is likely to get 3–4 eggs at most, and her chance of a live birth is no more than 2%–3% (7).

Poppy and Roger need to know all this before they go for IVF. They must think about whether they will attempt IVF with her own eggs or accept that they should think about using donated eggs. The FSH tests and others all point to incipient ovarian failure, and her follicle response at IVF will be greatly diminished, as the eggs (and follicles) are likely to be of poor quality.

What if Poppy hadn't been so pushy and insisted on blood tests when she saw the GP/primary care doctor? Should the GP have been more aware of the issues and paid attention to Poppy's 'ovulation kit', which suggested a problem? GPs encounter demanding patients quite often, patients who come with information from the internet and all manner of tests and remedies they have acquired over the counter. The GP needs incredible discernment (born out of knowledge and experience) to detect those patients who have something seriously wrong. No GP is going to rush to think of early menopause when a woman has been trying to conceive for only 9 months, but an inquiring attitude might have wondered why this woman developed irregular periods over the previous 2 years after previously regular cycles. GPs should know that one of the commonest causes is PCOS, a condition with long-term morbidity from type 2 diabetes and cardiovascular disease, and frequently identified or at least postulated in primary care. It might be expected that the GP would pursue this line of thought and request FSH, LH and oestrogen levels on the appropriate days of the menstrual cycle. Poppy shouldn't have had to insist on these tests.

3.10 SUMMARY

When commencing investigations of infertility, consider the main diagnostic categories and work through those simultaneously. Bear in mind that common things occur commonly (polycystic ovary syndrome and Chlamydia damage do occur commonly, but azoospermia and congenital absence of the vas deferens do not), but the patient's history will almost always give you, the clinician, what you need to direct you. The investigations in Table 3.1 will provide you with a firm basis for starting the process, and I would expect that they will be sufficient for 97% of patients you encounter.

REFERENCES

1. Corrigan E. Nurses as equals in the multidisciplinary team. *Human Fertility (Cambridge, England).* 2002;5(1 Suppl):S37–40.
2. National Collaborating Centre for Women's and Children's Health. *Fertility Problems: Assessment and Treatment.* London: NICE; 2013, p. 51.
3. Egbase PE, Al-Sharhan M, Grudzinskas JG. Influence of position and length of uterus on implantation and clinical pregnancy rates in IVF and embryo transfer treatment cycles. *Human Reproduction (Oxford, England).* 2000;15(9):1943–6.
4. Rushton DH, Ramsay ID, Gilkes JJ, Norris MJ. Ferritin and fertility. *Lancet (London, England).* 1991;337(8756):1554.
5. Akande VA, Hunt LP, Cahill DJ, Caul EO, Ford WCL, Jenkins JM. Tubal damage in infertile women: Prediction using Chlamydia serology. *Human Reproduction.* 2003;18(9):1841–7.
6. Westrom L. Effect of acute pelvic inflammatory disease on fertility. *American Journal of Obstetrics and Gynecology.* 1975;121(5):707–13.
7. Hull MG, Cahill DJ. Female infertility. *Endocrinology and Metabolism Clinics of North America.* 1998;27(4):851–76.
8. Cahill DJ, Prosser CJ, Wardle PG, Ford WC, Hull MG. Relative influence of serum follicle stimulating hormone, age and other factors on ovarian response to gonadotrophin stimulation. *British Journal of Obstetrics and Gynaecology.* 1994;101(11):999–1002.
9. Rotterdam ESHRE/ASRM-Sponsored PCOS Consensus Workshop Group Revised 2003 consensus on diagnostic criteria and long-term health risks related to polycystic ovary syndrome. *Fertility and Sterility.* 2004;81(1):19–25.
10. Meczekalski B, Czyzyk A, Kunicki M, Podfigurna-Stopa A, Plociennik L, Jakiel G et al. Fertility in women of late reproductive age: The role of serum anti-Mullerian hormone (AMH) levels in its assessment. *Journal of Endocrinological Investigation.* 2016;39(11):1259–65.
11. Nelson SM, Fleming R, Gaudoin M, Choi B, Santo-Domingo K, Yao M. Antimullerian hormone levels and antral follicle count as prognostic indicators in a personalized prediction model of live birth. *Fertility and Sterility.* 2015;104(2):325–32.
12. Hussain M, Cahill D, Akande V, Gordon U. Discrepancies between Antimullerian hormone and follicle stimulating hormone in assisted reproduction. *Obstetrics and Gynecology International.* 2013;2013:383278.
13. Steiner AZ, Pritchard D, Stanczyk FZ, Kesner JS, Meadows JW, Herring AH et al. Association between biomarkers of ovarian reserve and infertility among older women of reproductive age. *JAMA.* 2017;318(14):1367–76.
14. Lehman R. Hormone tests don't predict fertility in premenopausal women. *BMJ: British Medical Journal.* 2017;359(8216):115.
15. Glazener CM, Ford WC, Hull MG. The prognostic power of the post-coital test for natural conception depends on duration of infertility. *Human Reproduction (Oxford, England).* 2000;15(9):1953–7.
16. Nahum R, Shifren JL, Chang Y, Leykin L, Isaacson K, Toth TL. Antral follicle assessment as a tool for predicting outcome in IVF – Is it a better predictor than age and FSH? *Journal of Assisted Reproduction and Genetics.* 2001;18(3):151–5.
17. Luciano DE, Exacoustos C, Luciano AA. Contrast ultrasonography for tubal patency. *Journal of Minimally Invasive Gynecology.* 2014;21(6):994–8.
18. Akande VA, Hunt LP, Cahill DJ, Jenkins JM. Differences in time to natural conception between women with unexplained infertility and infertile women with minor endometriosis. *Human Reproduction (Oxford, England).* 2004;19(1):96–103.
19. Mills MS, Eddowes HA, Cahill DJ, Fahy UM, Abuzeid MI, McDermott A et al. A prospective controlled study of *in-vitro* fertilization, gamete intra-fallopian transfer and intrauterine insemination combined with superovulation. *Human Reproduction (Oxford, England).* 1992;7(4):490–4.

20. Barnhart K, Dunsmoor-Su R, Coutifaris C. Effect of endometriosis on *in vitro* fertilization. *Fertility and Sterility*. 2002;77(6):1148–55.
21. Cahill DJ, Wardle PG, Maile LA, Harlow CR, Hull MG. Pituitary-ovarian dysfunction as a cause for endometriosis-associated and unexplained infertility. *Human Reproduction (Oxford, England)*. 1995;10(12):3142–46.
22. Hull MG, Glazener CM, Kelly NJ, Conway DI, Foster PA, Hinton RA et al. Population study of causes, treatment, and outcome of infertility. *British Medical Journal (Clinical Research Ed)*. 1985;291(6510):1693–7.
23. Chambers GM, Hoang VP, Sullivan EA, Chapman MG, Ishihara O, Zegers-Hochschild F et al. The impact of consumer affordability on access to assisted reproductive technologies and embryo transfer practices: An international analysis. *Fertility and Sterility*. 2014;101(1):191–8.e4.
24. Boivin J, Griffiths E, Venetis CA. Emotional distress in infertile women and failure of assisted reproductive technologies: Meta-analysis of prospective psychosocial studies. *BMJ (Clinical Research Ed)*. 2011;342:d223.
25. Hyland A, Piazza K, Hovey KM, Tindle HA, Manson JE, Messina C et al. Associations between lifetime tobacco exposure with infertility and age at natural menopause: the Women's Health Initiative Observational Study. *Tobacco control*. 2016;25(6):706–14.

Influencing the sperm count 4

OVERVIEW

This chapter covers a basic understanding of sperm sample results, how to compare one sample result with another, a simple description of the way hormonal stimuli work together for sperm production, a description of sperm maturation and transport and clinical problems therein, clinical problems in sperm production, medical treatments to improve sperm counts and, finally, tips on overcoming male reluctance to have a sperm count.

The term 'male infertility' is used sparingly throughout this book. It is not a concept I subscribe to, thinking it to be too broad, ranging from azoospermia to erectile dysfunction. There are several sperm disorders, and the term male infertility promotes an approach of lumping all cases together without further evaluation. I prefer the term 'sperm dysfunction', as it is clearer and more meaningful.

4.1 SEMEN ANALYSIS: DESCRIPTORS AND METHODS OF COMPARISON

The most recent World Health Organization (WHO) criteria for semen analysis were published in 2010 (Table 4.1) (1). Semen analysis yields information on sperm concentration, motility and normality. Each parameter has value and importance, but, if a series of semen analyses are performed in the same man, each parameter will show variation over time, making it difficult to compare one result with another and to interpret the results.

One solution is to use an approach developed in Bristol by my predecessor, Michael Hull. This is the motile normal sperm density (or sperm count), the MNSD(C). It is a simple mathematical calculation that allows a clinician to amalgamate the three parameters into one, allowing for comparison between one sample and another.

It is calculated like this: $\text{MNSD(C)} = \text{Concentration} \times \text{Motility} \times \text{Normality}$

For someone with a sperm concentration of 50 million/mL, motility of 40% and normality of 10%,

$$\begin{aligned}\text{MNSD(C)} &= \text{Count} \times \text{Motility} \times \text{Normality}\\ &= 50 \times 40/100 \times 10/100\\ &= 50 \times 0.4 \times 0.1\\ &= 2 \text{ million motile normal sperm in the sample}\end{aligned}$$

The MNSD(C) is applied in Table 4.2 which presents one man's hypothetical results over several months. It may be obvious that the April 2017 values are the worst of the three, but which is the better of the other two? Using the MNSD(C) calculation, the March 2018 is the better of the two samples. That is clearly important in assessing whether there has been change in response to an intervention, such as smoking cessation.

4.1.1 Sperm quality

The sperm count is a test of sperm numbers only – it tells little about the quality of the sperm sample and how good the sperm are at fertilising eggs. There are multiple tests of sperm function currently available, which include tests of the acrosome reaction, reactive oxygen species, sperm DNA integrity, sperm penetration assays and the recovery and survival test. Of these, the recovery of sperm after swim-up into culture fluid, and their subsequent 24-hour survival, the *recovery*

Table 4.1 WHO sperm characteristics.

Parameter	WHO 2010
Volume	at least 1.5 mL
Concentration	at least 15 million/mL
Total number	39 million in the ejaculate
Progressive motility	at least 32%
Total motility (progressive and nonprogressive motility)	at least 40%
Normal forms	at least 4%
Vitality: Live, immotile versus dead spermatozoa	at least 58%

Source: Adapted from Cooper TG et al. *Human Reproduction Update.* 2010;16(3):231–45.

Table 4.2 Semen analyses over time in one man.

	Sample 1	Sample 2	Sample 3
Date	Oct 2016	April 2017	March 2018
Concentration ($\times 10^6$/mL)	35	15	28
Motility (%)	40	20	55
Normality (%)	10	15	35
MNSD(C)	1.4	0.45	5.4

and survival test, is an easy sperm quality test to undertake and it has predictive value. Swim-up describes a technique in which semen are placed in a test tube and a culture medium is inserted on top of the semen. Under centrifugation, healthy sperm will swim up into the culture medium, leaving debris and slower or immotile sperm behind. This test helps in identifying those men likely to father a child naturally, and those likely to require assisted conception techniques such as intra-uterine insemination, intracytoplasmic sperm injection or IVF. It can also predict successful fertilisation rates: 48.3% fertilisation rate if spermatozoa recovered are $\leq 2.0 \times 10^6$ versus a 71.4% rate if recovery numbers are $>2.0 \times 10^6$ (2).

There are other techniques available, designed to look at sperm characteristics in greater detail. These computer-assisted sperm analyses (CASA) use computer software to recognise sperm shape and movement, replacing the human assessor. The technique has been developed over the past 30 years but has yet to be taken up by many reproductive medicine clinics, mostly because the software is under continuous improvement and upgrading. It is the subject of research papers and the testing is used in research laboratories and private clinics. A 2017 study in Poland, comparing CASA methods with manual assessment found that CASA underperformed; their conclusions reflect my view that CASA is, as yet, not ready for adoption by mainline services (44). There are, however, data to support the value of some CASA parameters in predicting fertility in men with normal infertility.

Separately, DNA fragmentation in sperm is being intensively investigated, and a comprehensive review is recommended for more details (45). Sheena Lewis in Belfast, Northern Ireland, argues that sperm DNA analysis should be a routine part of sperm assessment as it is a good indicator of sperm quality. Currently, this test is not available in public clinics unless the patient is able to pay for it separately.

4.1.2 Obtaining a sperm sample

Producing a sperm sample is a difficult task for many men. Some men will never have masturbated. Of those who have, some will find it difficult to do so to order. Tired of repeatedly instructing men how to masturbate so as to get the specimen into the collection bottle in the desired manner, some years ago I produced a video clip, now available on YouTube, which explains the technique and correct timing required for good sample production (https://www.youtube.com/watch?v=5GTfSQxz9s4).

If the issue is one of distaste for 'performing' in a clinical environment, or if there an ideological objection to masturbation, the following suggestions may help (in order of increasing difficulty):

- Enquire whether producing the sample at home would make it easier for the couple.
- Use a specialised collection condom for semen analysis and fertility treatment.
- Use an approved lubricant which is not spermicidal, if masturbation without it proves difficult.
- Very rarely and as an absolute last resort (in an emergency on the day of egg collection), I have seen electro-ejaculation or surgical sperm retrieval used to retrieve sperm.

Some men, however, may be uncooperative about having a sperm test because they cannot accept that the couple's difficulty conceiving has anything to do with them. Unfortunately, there are no easy solutions to this psychological block.

Essentials for a good sperm sample:

- Appropriate interval between last ejaculation and sample collection (at least 48 hours and no more than 7 days).
- Getting all the sample in the pot (especially the first 'spurt', as this contains the highest sperm numbers [3]).
- Avoiding contamination of the penis with bacteria (by washing), to avoid confusion at analysis between sperm cells and bacteria.

4.2 CAUSES OF LOW VALUES ON SEMEN ANALYSIS

4.2.1 Hypothalamic-pituitary disorders

Nonhormonal space-occupying pituitary tumours which impinge on the areas producing gonadotrophin hormones. Prolactin-secreting tumours in men will usually present with direct evidence of the condition (galactorrhoea or visual disorders), but of course may present only with absent or very few sperm.

Hypogonadotrophic hypogonadism (HH). There are two types – a pubertal type characterised by abnormal pubertal development, infertility and low gonadotropin levels, and an adult-onset HH in men who have been through normal puberty. The pubertal type includes such rare conditions as Kallmann syndrome (failure of GnRH production and anosmia). One cause of adult-onset HH is panhypopituitarism as a result of radiotherapy or surgery for a brain tumour. The patient is left with complete absence of pituitary function, requiring replacement of all the pituitary stimulated hormones (thyroxine, cortisol, testosterone, growth hormone).

4.2.2 Gonadal failure

Primary causes such as failure of the testes to descend or chromosomal causes (see Section 4.2.4.1). Secondary causes such as orchitis, trauma or drugs.

4.2.3 Normal variability

Semen analysis results vary considerably in men from day to day and week to week. The best evidence for this comes from men donating sperm – there can be huge variation in sperm count values in any given man, some have sperm results ranging from less than 10 million/mL to over 200 million/mL (4).

4.2.4 Causes related to sperm transport (see Section 4.6.2)

- Abnormalities of the vas deferens
- Ejaculatory dysfunction (retrograde ejaculation, for instance)
- Spinal cord damage from trauma
- Vascular disorders and diabetes

4.2.5 Chromosomal abnormalities – Y chromosome deletions

If oligozoospermia or azoospermia is present and investigation excludes endocrine and obstructive pathology, then chromosomal causes should be considered. When sperm counts are less than

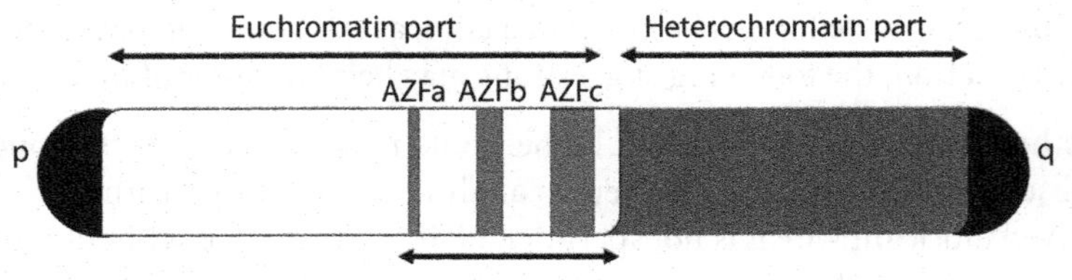

Figure 4.1 Y-micro deletions (AZF, azoospermia factors).

1 million sperm/mL, and in practice, at least 1×10^{-3} million sperm/mL or less, men should have detailed examination of the Y chromosome, looking for microdeletions on the long arm of the chromosome (5) which are known to cause very low or absent sperm numbers. Those deletions are illustrated in Figure 4.1, and the pattern and number of deletions determine whether sperm are present, either at all or in very low numbers. A fuller account of AZF deletions on the Y chromosome (and the only other description of the condition), is given the first 'Case Presentation' at the end of this chapter.

4.3 HORMONAL STIMULI LEADING TO SPERM PRODUCTION

Spermatogenesis is the result of a delicate interplay between pituitary hormones, endocrine activity in the testis and induction of receptors in the cells of the testis. In an endocrinologically competent man, two hormones, FSH and LH, work together to stimulate the production of testosterone and sperm (Figure 4.2). (LH in men was previously called interstitial cell stimulating hormone). The endocrine control of spermatogenesis is illustrated in Figure 4.3a. GnRH signals the anterior pituitary gonadotroph cells to produce FSH and LH. LH binds to receptors on the surface of the Leydig cells (previously called interstitial cells) to induce production of testosterone. Testosterone moves by diffusion into capillaries, and thence into the circulation to have endocrine effects elsewhere (muscle, bone, etc.). Locally, it moves across the basement membrane into adjacent Sertoli cells and seminiferous tubules to have a paracrine effect. This is a key step and must be mimicked if exogenous gonadotrophin treatment is used therapeutically. Sertoli cells possess receptors for testosterone and FSH. Testosterone induces the receptors which respond to FSH (46). The basement membrane, illustrated in Figure 4.2 as a thick line separating the Leydig cells and vasculature from the Sertoli cells is clinically and immunologically important. It serves as a barrier, the 'blood-testis' barrier, keeping the immunologically different spermatozoa separate from the rest of the body. Full details of intracellular changes and signalling can be found elsewhere (47).

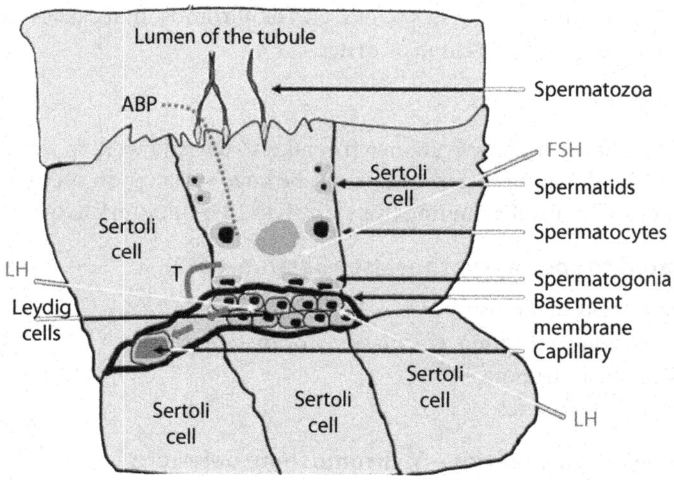

Figure 4.2 A section across a seminal vesicle (structures in black/white; hormones in grey).

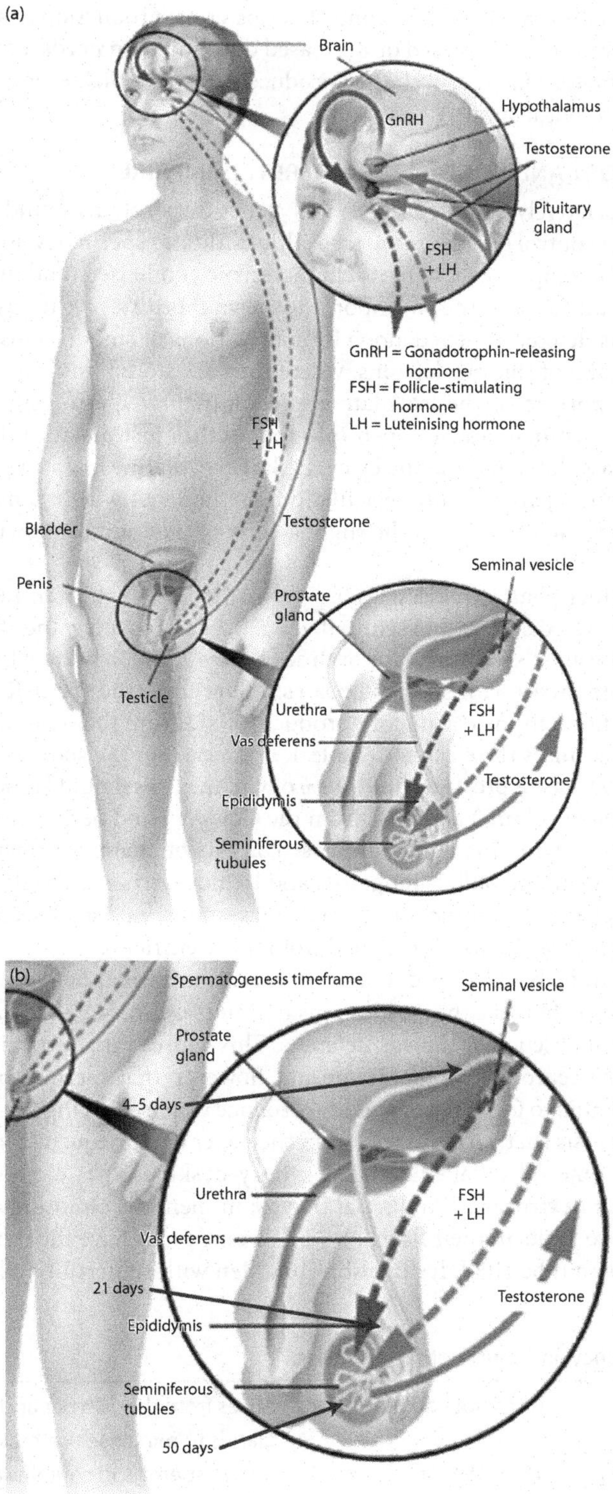

Figure 4.3 (a) The endocrine control of spermatogenesis; and (b) the timeframe for the passage of sperm from production in the testis to ejaculation. (From Wardle PG, Cahill DJ. *Understanding Infertility*. Family Doctor Publications, Poole, United Kingdom; 2005, with permission.)

In terms of hypothalamic and pituitary control, testosterone production from the Leydig cells in the testis cause LH levels to be increased or decreased depending on needs; Sertoli cells respond to FSH to produce spermatozoa and as well they produce inhibins which exert negative feedback on the pituitary and the hypothalamus.

4.4 SPERM TRANSPORT AND CONDITIONS WHERE IT IS IMPAIRED

The epididymis is a relatively complex functional structure. Proteins found in epididymal fluid work with enzymes to control the sperm surface. This facilitates sperm development from a round germ cell into a highly complex cell, with its characteristic condensed (and inactive) nuclear head and the flagella-like tail to assist the so-important sperm motility. Sperm progress from the epididymis along the vas deferens, moved along by contractions of smooth muscle surrounding the vas to reach the ampulla of the vas deferens, where they are stored until the next ejaculation. At ejaculation, the first portion of the ejaculate is predominantly sperm cells and prostatic fluid, which is rich in zinc. Seminal vesicles open into the urethra just distal to the vas deferens, and the fluid is generally alkaline, making the ejaculate largely alkaline in nature (pH ranges between 7.35–7.50) (7). It becomes jelly like after ejaculation and liquefies within 5 minutes normally (8). The alkalinity may serve to balance out the slight acidity of the vagina, allowing sperm to survive slightly longer.

There are two distinct phases in ejaculation – emission and expulsion. Emission involves the distal epididymis, the vas deferens, the seminal vesicles, the prostate gland, the prostatic urethra and the bladder neck (most structures are outlined in Figure 4.3b). After bladder neck closure, fluid from the prostate mixes with spermatozoa-rich fluid from the vas deferens in the prostatic urethra. The final addition to the ejaculatory product is fluid from the seminal vesicle, which contains fructose and alkalinises the ejaculate. Table 4.3 summarises the source of the components of the ejaculate which consists of prostatic fluid (10% of volume), vasal fluid (10% of volume), seminal vesicle fluid (75%–80% of volume) and fluid from the Cowper's and periurethral glands (or glands of Littre). Expulsion following the emission phase consists of clonic contractions of the prostate and several muscles (including bulbospongiosus and ischiocavernosus, levator ani and transverse perineal muscles). Each clonic contraction lasts 0.6-1.0 seconds, with a 0.7 second interval between contractions, and in this way, semen is expelled from the urethra (6).

The vas deferens can be blocked or damaged in several unrelated medical conditions. In male children born with cystic fibrosis, the vas deferens is almost uniformly absent, a congenital condition involving bilateral absence of the vas deferens. This causes difficulty with fertility from the beginning of the man's reproductive life. It appears, however, that sperm aspirated from the epididymis have the potential to fertilise oocytes and produce viable offspring (9) – whether these were screened for cystic fibrosis isn't entirely clear from the paper or subsequent correspondence (9).

Kartagener's syndrome (also called primary ciliary dyskinesia) is a condition characterised by bronchiectasis with dextrocardia and nasal polyps; in men it is almost always associated with immotile sperm (due to faults in their flagella), while women may have tubal infertility considered to be related to fallopian tube cilial dysfunction (10). Men with subfertility related to Kartagener's

Table 4.3 Constituency and source of the ejaculate.

Pre-ejaculatory phase	Does not contain sperm	Arises from the Cowper and Littre's glands to minimise urethral acidity
First ejaculated fraction	15 to 45% of the whole ejaculate	Rich in sperm and contains epididymal and prostatic secretions, rich in acid phosphatase, citric acid, magnesium, and zinc
Second ejaculated fraction	Remaining 55 to 85% of volume	Low sperm count secretions from the seminal vesicles

syndrome are not able to conceive naturally. Some reports using testicular sperm or ejaculated sperm (and ISCI) reported fertilisation rates of 65% and 55% (no significant differences noted) (11).

4.5 CHEMICAL INHIBITORS AND LIFESTYLE CHOICES: EFFECT ON SPERM PRODUCTION

Sperm production is a delicate process and is easily thrown out of kilter so that it fails or is significantly suppressed. I frequently see men in clinic who have been body building and taking supplements to assist in building up muscle mass, but when sperm counts are done, the man has little or no sperm. Not surprisingly, it causes consternation to the couple when the explanation given is of sex steroid induced subfertility. Men tend to be unaware that these supplements contain substances that can diminish the sperm count (28). People generally expect that increasing one's testosterone levels would improve sperm count, but the reverse is true, and the process of recovery takes several months after anabolic steroids/supplements are stopped. It takes at least 4 weeks for washout of the anabolic steroid to occur (12), followed by the time needed for spermatogenesis to produce fully mature spermatozoa (see timeframes in Figure 4.3b).

There are particular occupational exposures which can also alter sperm production. Men working in high-temperature environments (glass works, foundries, mines, bakeries and laundries, as well as occupations which require heavy-duty protective clothing) will experience some degree of negative effect on sperm production (13). Past research from this hospital, working with the ALSPAC data, showed that men working in the printing industry take significantly longer to conceive (14). Others have shown that men working in toll booths on the Italian Autostrada have higher than expected levels of subfertility with significantly lower values for their sperm counts compared with matched controls, raising questions about toxic chemicals from vehicle exhaust fumes (15).

There is ample evidence which supports the negative effect of cigarette smoking on fertility, and particularly on sperm production. The data show that cigarette smoking suppresses sperm quality across the parameters of concentration, motility and normality in a dose-dependent manner. This is related to reduced levels of testosterone and FSH in smokers, as well as the detrimental effect of other factors such as the amount of reactive oxygen species in the ejaculate. Fortunately, smoking cessation leads to a reversal of these negative findings after 3 months of cessation.

Does mobile phone use affect sperm mobility? That question was irrelevant 20 years ago, but now almost every person walking around has a radiofrequency electromagnetic radiation emitter (a mobile phone) in their pocket. Animal experiments with controlled exposure to mobile phones suggest decreases in sperm count and motility and increases in oxidative stress (16). A meta-analysis of human studies showed sperm motility to be approximately 8% lower in exposed compared with nonexposed groups (17). In itself, this might be considered a small effect, but if we remember that infertility is generally multifactorial and that men are constantly exposed to multiple environmental insults, it does have relevance. Continuing the technological theme, an *in vitro* study with divided sperm aliquots had one-half allocated to Wi-Fi exposure from a laptop, the other half not. The exposed group had a decrease in sperm motility and an increase in sperm DNA fragmentation (18). The question is, how many men would be prepared to forgo a mobile phone (or have it switched off unless required) or to give up using a laptop?

Laboratory research suggests that reactive oxygen species (ROS) and oxidative stress are related to poor sperm function and to their ability to fertilise. Researchers in Bristol found many years ago that the presence of ROS is linked to poor sperm motility and to the ability of sperm to undergo the acrosome reaction, a key stage in preparing the sperm to penetrate an oocyte (19). Turning that knowledge into practical solutions that can improve sperm parameters has so far proven difficult. Several years ago, a review of randomised controlled studies using over-the-counter vitamins and supplements available in the United States over the last 3 decades was unable to draw any conclusions because of inconsistent and poor-quality data (20). Some underpowered studies have shown the effect of antioxidant therapy such as selenium and Vitamin E to be positive (21,22), but a randomised multicentre study failed to show any difference in sperm quality of men either treated or untreated with antioxidants (23). So, this question is still unresolved but in my view, the evidence is not supportive of antioxidant use.

High alcohol use is generally accepted to have a negative effect on male fertility, and a detailed alcohol intake history should be taken if male infertility is suspected. Stopping or reducing alcohol may lead to improved sperm numbers and function but low and moderate intake has repeatedly been shown not to affect a man's ability to achieve a pregnancy (24–26).

Most recreational drugs are associated with reduction in sperm quality, but the best evidence is for cannabis use. Marijuana has two active cannabinoids, delta-8-TetraHydroCannabinol (THC) and delta-9-THC, which inhibit mitochondrial oxygen consumption, leading to disorders of sperm respiration function (27).

Most exercise appears not to have any effect on sperm quality apart from excessive (more than 5 hours a week) cycling, probably a consequence of hyperthermia in the scrotum (27). What may be surprising is that the definition of excessive cycling in that study was more than 5 hours a week; people who use a bicycle as a regular means of transport for commuting can easily exceed this figure in the course of a normal week. A literature review found evidence that long-distance cycling affects erectile ability, mediated through perineal compression of the pudendal nerve within Alcock's canal, and linked to saddle shape (31). However, this view is not supported by a survey of 4000 self-selected survey respondents, which concluded that cyclists were no more likely to experience sexual or urinary dysfunction than swimmers or runners (32). Intensive endurance training seems to have the effect of reducing testosterone by poorly understood mechanisms, reducing the drive to sperm production (29,30).

As a convenient summary, NICE guidelines on lifestyle effects on sperm quality (33) give the following recommendations (with guideline references from the NICE guidelines in parentheses):

1. Three to four units of alcohol per day for men are unlikely to affect their semen quality (1.2.3.2).
2. Excessive alcohol intake is detrimental to semen quality (1.2.3.3).
3. Smoking is associated with reduced semen quality and stopping smoking will improve their general health (1.2.4.4).
4. A BMI of 30 or over is likely to lead to reduced fertility (1.2.6.4) (though evidence is lacking for this statement).
5. An association does exist between elevated scrotal temperature and reduced semen quality, but there is no evidence that wearing loose-fitting underwear improves fertility (1.6.30).
6. Prescription, over-the-counter and recreational drugs can interfere with both male and female fertility; enquiry about these should be made with people who are concerned about their fertility, and appropriate advice should be offered (1.2.10.1).

4.6 TREATMENTS

4.6.1 Medical treatments for low or absent sperm numbers in association with other endocrine disorders

Treatments have a better chance of success if used for the right condition! Failing to resist pressure from the emotional plight of the couple in front of you can lead to unsuccessful treatment which is resource intensive, both physically and financially, and to disappointment for the couple.

Hypogonadotrophic hypogonadism or failure to produce sperm is due to impaired gonadotrophin release: the man has little or no sperm, with low levels of the gonadotrophins, FSH and LH, and low levels of testosterone. Ideally, the treatment plan will follow a 'recipe' such as that in Figure 4.4, adapted from a large multicentre study showing its efficacy (34). Treatment is a long process, taking up to 6 months. Any exogenous testosterone prescribed as a hormone replacement should be stopped, allowing 4 weeks for it to wash out and dislodge from the receptors in the testis. The next step is to raise endogenous testosterone levels above a key threshold (which we have set at 10 nmol/L) using human chorionic gonadotrophin (hCG), which is analogous to LH or luteinising hormone (details of doses for hCG and FSH are provided in Figure 4.4). Once that threshold of testosterone is achieved, FSH is started. It takes at least 3 months for sperm to appear in the ejaculate from that time, and testing for spermatozoa after 4 months is advisable. If no sperm or low levels of sperm are found in the sample, increase the dose of FSH and then retest 3 months later. If, at the end of the second phase of treatment, there are still no sperm in the sample, treatment should end on pragmatic and financial grounds. By the end of the second phase of treatment, a conservative estimate of the drug cost is about £1500.

Induction of spermatogenesis: Male infertility due to hypothalamic or pituitary failure

Setting	Reproductive Medicine/IUI clinic
For Staff	Staff within the Reproductive Medicine Department
Patients	Patients receiving Induction of spermatogenesis

Suitable for men with low gonadotrophin secretion due to pituitary surgery, pituitary tumours, hypothalamic disorders.

Expected clinical and laboratory findings:

- Relevant past history
- Small/atrophic testes
- Low circulating testosterone levels (<10 nmol/L) though in practice most men will be on HRT such as *Sustranon* injections

Management

Immediate

- Semen analysis as a baseline before hCG injections
- Check female baseline tests (FSH/LH, TSH, Chlamydia, Rubella, Hepatitis B & C)
- Stop any hormone replacement therapy

1. Start hCG: IM injection of hCG (2000 iu) 3 times weekly (Mon, Wed, Fri)
2. At 4 and 8 weeks, check serum testosterone
3. Once testosterone concentration exceeds 10 nmol/L, continue hCG and only discontinue when FSH/hCG is discontinued; go to number 5
4. If testosterone concentration ≥ 10 nmol/L, increase hCG to 3000 iu (or if necessary 4500 iu a month later)
5. Add in FSH/hMG 75 iu (three times weekly) for 4 months
6. If spermatogenesis has not ensued, increase to 150 iu (three times weekly) for 4 months

Monitor: S. testosterone
Testicular size
Semen analysis

Long-term

It is not feasible for us to maintain men on Gonadotrophin therapy until they achieve a pregnancy once spermatogenesis has begun. Therefore, we need to stipulate in advance that once spermatogenesis has begun, we can only maintain Gn for three months to allow these patients to store sufficient quantities of sperm so that they can use these to achieve pregnancy by cervical intrauterine insemination or by IVF or ICSI. They may have to pay for this themselves.

QUERIES

Figure 4.4 Induction of spermatogenesis.

Bristol experience with this treatment regime is positive, and, provided the diagnosis is correct, the likelihood of getting viable, usable sperm is high (13 out of 16 men, just over 80%) (35).

Once spermatogenesis has started, the administration of LH alone will allow sperm production to continue for a long time, although the quality of the sperm will diminish over 12 months (36). Consequently, concerted efforts to achieve conception and to cryopreserve sperm should be made once sperm production is achieved.

Be aware when managing hypogonadotropic hypogonadism caused by panhypopituitarism, that these men require growth hormone (GH) supplementation, or an increase in the already prescribed dose of GH in order to achieve successful spermatogenesis.

Unfortunately, most other causes of azoospermia and oligozoospermia are not responsive to medical therapy.

4.6.2 Surgical interventions in the case of low or absent sperm counts

Even when sperm numbers are very low or absent, it may be possible for sperm to be obtained by surgical exploration of the testis. Sperm may be found in the testes if the man has normal gonadotrophin levels and normal testicular volume.

Such men should be offered an intervention to recover sperm and store whatever is not immediately required in case of future need. This can be done surgically by needle epididymal aspiration or by testicular biopsy with exploration of the biopsied material. Indications for epididymal exploration are that the epididymus can be felt on palpation. This approach is less traumatic as it is done by needle aspiration, and the advantage is that one is more likely to find sperm, since this is where maturing sperm are stored.

If the absence of sperm is due to ejaculatory failure because of a spinal injury, or if there is neurological damage caused by vascular disease, diabetes or multiple sclerosis, then ejaculation can be artificially induced using electrical stimulation to the prostate and the seminal vesicles (electro-ejaculation). Men with spinal cord lesions will not require anaesthesia, while all others will. A rare complication of electro-ejaculation is autonomic dysreflexia, characterised by severe hypertension, occurring even under anaesthesia (but rare in men whose lesion is below the sixth thoracic vertebra). If this occurs, the procedure should be stopped and antihypertensives administered. Otherwise the procedure is relatively free of complications. Failure to obtain sperm from these men is uncommon – fewer than 10% in my experience. The technique is not widely practiced as it can be distasteful to clinical staff because of the intimate nature of the procedure.

4.7 OVERCOMING MALE RELUCTANCE/REFUSAL TO HAVE A SPERM COUNT

In many cultures (in the developed and developing world), it is not uncommon for the male partner to struggle with the challenges of being investigated for infertility, and an incorrect association is made between sperm disorders and virility or masculinity. The ability to produce children strikes at the heart of a man's identity, and feelings of embarrassment, shame and failure come to the surface when this is questioned. The male peer group is frequently to blame for this, as they make 'humorous' comments on people's sperm numbers which are hurtful, damaging and untrue.

For whatever reason, it happens that men will sometimes not cooperate with their female partner, and when this occurs, ways to bypass the problem must be sought. One option is to use a range of branded spermicide-free condoms (readily available on the internet) for use in vaginal intercourse, and the sample can be collected from this (using these does mean that care needs to be taken to remove the condom after ejaculation and before detumescence occurs, the condom needs to be sealed and then transported within an hour to the clinic for analysis). Another possibility is to use the postcoital test, which means the couple have sexual intercourse, after which the woman presents herself to the clinic to have an endocervical mucus collection performed, which will capture her partner's sperm. This is less demanding on resources in one way, but does require trained staff in a clinic to be able to accurately interpret a sample. Full details of this test are published elsewhere (37), but a broad summary is provided here. This test is undertaken by asking the woman to have intercourse between 14 and 16 days before menstruation is due, and the test is done 12–18 hours later (so having intercourse the night before and attending in the morning is very feasible). Using a speculum, the cervix is exposed and a 1 mL (10 cm-long) plastic syringe collects endocervical mucus by placing the tip into the cervical canal. Light suction will draw mucus up into the syringe, and it is then transferred to a microscope slide for examination at high (400×) magnification. A count is made of the number of forward-progressing sperm in each high-powered field. Results range from negative (no sperm seen) through fair (1–5 sperm per high-powered field) to good (more than 20 per high-powered field). These results correlate well with the likelihood of natural conception (38), while not giving a precise estimation of actual sperm numbers. Most clinics have stopped using this test for a multitude of reasons, partly because results were variable and lacked reliability (39) and partly because assisted reproduction techniques have taken away the need for extensive testing of sperm or sperm function. We have continued to use the test judiciously because it is easy to do, has no added costs and gives good prediction of the likelihood of natural conception provided the couple have been trying for less than 3 years (38).

4.8 CASE PRESENTATIONS AND COMMENTARY

4.8.1 Case one

Mike Carrott is a fit 32-year-old man who exercises once or twice a week. He weighs 80 kg and doesn't smoke or drink alcohol. He has been married to 31 year old Sue for 3 years, and they have been attempting to have a child for the last 2 years. They decided to seek help through their GP. Mike and Sue made an appointment with their family doctor to discuss their failing to conceive. The doctor agreed that they needed to see a specialist and referred them. Meanwhile, several blood tests for Sue and a sperm test for Mike were arranged. The sperm test was arranged quite quickly, and the result came back to the family doctor. The results showed there were no sperm, and the doctor rang the couple to ask them to attend to discuss this. Both Mike and Sue were very upset, realising how serious a problem this was.

They saw the specialist about 4 weeks later. She discussed their history, drugs, operations, smoking and alcohol, and then asked to examine them both. In the history, she noted that Sue had two sisters, one younger and one older, neither of whom had tried to have a child. Mike had two brothers. He is the youngest of three boys and neither of the other brothers have children, despite trying for 3 and 4 years, respectively.

Sue had a transvaginal scan, allowing the specialist to see that her womb looked normal and her ovaries were normal in appearance with a small number of follicles and one large one measuring 15 mm on the right ovary. Mike had a normal scrotum, with no gross abnormalities. His right and left testes and epididymes were palpated; the testes were 20 mL in volume, and a vas deferens wasn't palpable on either side.

The specialist discussed these findings with Mike and Sue, and arranged some hormone assays (FSH, LH and testosterone) and genetic blood tests (Y chromosome deletions and screening for cystic fibrosis) for Mike, to investigate and assess him further.

At review, 1 month later, the specialist had Mike's results to hand. All his hormone tests were normal, including FSH and testosterone. His results for cystic fibrosis were normal, (details in the Commentary below) but his Y chromosome deletions were abnormal, showing a deletion at the azoospermia factor C (AZFc) section of the chromosome (details in the 'Commentary' below). This was the likely cause of Mike's azoospermia.

This news was distressing for the couple, but the specialist pointed out that of the three types of Y chromosome deletion, this was probably the least pessimistic type to have. With this deletion, it was likely that there were sperm being produced in the testis, possibly even collecting in the epididymis, but the low numbers of sperm meant that there would not be any sperm in the ejaculate (40). Achieving conception would mean undergoing some sort of surgical procedure to remove the sperm from the epididymis or the testis. If sperm were obtained, these could be then stored in freeze storage until Sue and Mike were ready to have IVF with ICSI treatment. ICSI is the treatment of choice when the number of available sperm is low, numbering in the hundreds. The standard IVF techniques, to be successful, require numbers to be in the order of millions of sperm.

The specialist also said that it was likely that if there were male offspring resulting from Mike's sperm, that some or all of them could carry this genetic abnormality and experience azoospermia. Female offspring would not carry it or pass it on to any offspring. The specialist said that it was also possible that Mike's two brothers had this chromosomal abnormality and that it had come, most likely, from their father (in these cases, it is assumed that a *de novo* abnormality developed in the sperm of the father, not affecting his fertility).

Mike and Sue left the specialist clinic with mixed feelings. The potential was there for them to have their own biological baby, but they had to contend with the likelihood of that baby, if a boy, carrying the abnormality and the challenge of whether to tell his brothers about this news. They didn't talk about it for weeks, not knowing how to approach it. Eventually Sue broke the silence and the discussion started. They worked out that Mike would have the surgical operation, and if they got sperm they would continue with the IVF and ICSI procedure as soon as possible. They decided that if they got pregnant, they would then go and talk about it with his brothers.

Commentary: About 2% of men have severe oligospermia or azoospermia. Of the nonobstructive causes of oligospermia, up to 20% are due to deletions of AZFc (41). One study has examined the male offspring of men with AZFc who underwent IVF and ICSI – all the male offspring were found to carry the deletion, and are likely to experience severe oligozoospermia or azoospermia as adults, with consequent infertility (41). We should recognise the impact this may have on parents and the reluctance they might feel about putting a child through that experience, given that they themselves have just been through it.

All of Sue's results were essentially normal, with no significant findings that would impair their fertility as a couple. For Mike, two specific chromosome tests were undertaken: Examining for deletions of the Y chromosome and cystic fibrosis screening because of the clinical finding of the absence of the vas deferens.

The cystic fibrosis screening showed this result: This patient does *not* carry any of the cystic fibrosis transmembrane conductance regulator (CFTR) pathogenic variants listed, that is, it is normal for these 50 CFTR pathogenic variants and does not carry the partially penetrant CFTR intron 8 5T variant. This result significantly reduces the likelihood that this patient's azoospermia is CFTR related.

The Y chromosome deletion testing showed this result: This analysis indicated a complete deletion of the AZFc region. This deletion is likely to be the cause of this patient's azoospermia. The deletion does not include the distal Yq heterochromatin, that is, it is an interstitial deletion. Deletions of the AZFc region are associated with variable clinical and histological phenotypes. Sperm may be retrievable for ICSI. The likelihood of finding spermatozoa by testicular sperm extraction in a man with complete AZFc deletion is approximately 50% (40). Genetic counselling is recommended for this patient, especially prior to treatment with assisted reproduction techniques since this patient's AZFc deletion would be transmitted to any sons. Male relatives of this patient may wish to consider testing, due to the possibility of inheritance or germinal mosaicism for the deletion in the father.

The cystic fibrosis screening was negative; Mike is not a carrier for CF or a patient with azoospermia as the expression of the disease. The Y chromosome deletion is the likely cause of his azoospermia, and the diagnosis allows them to progress with the treatment plan as outlined above. The other two forms of Y deletion, AZFa and AZFb, have a much poorer prognosis in terms of retrieving sperm and fathering a child (42,43).

4.8.2 Case two

Bobbie (Roberta) and Billy Stone presented to their family doctor because they haven't been able to get pregnant, despite trying for over 2 years. In fact, it is even more serious than that, as Billy has been finding it difficult to get erections when they are having sex, and they are both distressed by this. The family doctor looked back through Billy's records and noted that he had an operation on his brain 3 years ago. He asked Billy about this. Billy recalled he was having problems with persistent headaches and found his job as an accountant to be more difficult, as he noted, when looking at spreadsheets on his desk, that only the middle of the page was visible. The sides disappeared when he looked at the centre of the page. Billy then volunteered that before the operation, he had occasionally noted some wetness around his nipples – he hadn't said this before, as he was embarrassed about it.

The family doctor referred the couple to see a specialist in the local hospital. She listened to their story, took a full history, and examined them both. Billy had a full range of vision and very little body hair, and his testicles were soft and 10 mm in diameter, with no surgical scars. Billy told her that the operation was done up through his nose, and he was bruised for weeks after. The specialist arranged for them both to have several blood tests – routine viral screening and hormone tests for both, and a sperm test for Billy. She arranged to see them again 2 months later.

The viral screening was negative, and Bobbie's hormone tests were unremarkable. The results indicated that she was clearly ovulating. Billy's tests were more challenging. His FSH and LH hormones were both less than 0.5 IU/L, and testosterone was 4 nmol/L. The specialist had arranged assays of his other pituitary hormones, and these were essentially normal. Semen analysis showed that there were no sperm in the ejaculate, despite centrifuging the sample to concentrate anything in the ejaculate.

The specialist evaluated the evidence together with the history. The operation previously performed on Billy was a transphenoidal hypophysectomy. She considered that much of his pituitary had been damaged as a result, which explained why the gonadotrophin hormones (FSH and LH) were very low. Because his LH was low, this led to testosterone being low and his semen analysis showing no sperm. She diagnosed hypogonadotrophic hypogonadism (lack of function of the gonads as a result of lack of stimulation by gonadotrophins) and discussed the treatment options. In particular, she discussed injectable treatments to produce sperm (Figure 4.4). She pointed out that the likelihood of producing sperm from this treatment was high. In a Bristol study (35), 80% of men on treatment were successful in producing sperm to be stored for treatment. Both Billy and Bobbie thought this was worth trying, and the specialist arranged an appointment with the nurse specialist who would manage the treatment and teach them how to do the injections. After 4 weeks of injections, Billy's testosterone was 15 nmol/L, and FSH injections were started. After 4 months on FSH injections, the first analysis of sperm showed several hundred motile sperm, and arrangements were made to freeze a few samples to allow further treatment by assisted conception. At their next visit, arranged to discuss assisted conception, Bobbie told the specialist that her period was late. After a pregnancy test, the specialist was able to tell them that she was pregnant! Despite this, they decided to go ahead and freeze sperm for possible future use.

Commentary: This is a story with several strands of fact woven in from real patient journeys. Twice in my career have I seen women conceive during the injection treatment phase in their male partners, and of course, for the couple, it is a source of great joy. But it is not the norm. Hypogonadotrophic hypogonadism, as this man had, is the condition most likely to respond to gonadotrophin therapy, especially when it is likely that he produced sperm before he had his pituitary tumour (35).

This is an interesting case because his presentation with infertility, some erectile dysfunction, and the smaller soft testicles on examination with very little body hair all pointed to a low testosterone level. The earlier presentation with headaches and visual disturbance (he had bitemporal hemianopia) is not uncommon for a nonsecreting pituitary tumour, but the breast leakage suggests a prolactin-secreting tumour.

He responded well and quickly to the injectable gonadotrophins. The normal expectation of this treatment would be to produce sufficient sperm to freeze several aliquots, to be used as part of assisted conception with intracytoplasmic sperm injection rather than ordinary *in vitro* fertilisation. It is vitally important to get the man's testosterone above the level of 10 nmol/L – without that, there will be no priming of the Sertoli cells by induction of receptors for FSH and consequently no spermatogenesis (34).

4.9 SUMMARY

Evaluation of a man's sperm is initially relatively simple, but the causes of abnormalities seen on an initial sperm sample may be difficult to treat. Considering the various hormonal/endocrine, mechanical and chromosomal/genetic causes is likely to be a useful approach in reaching a diagnosis.

The effect of men's reluctance to produce a sperm sample, and their discomfort in engaging with the idea that difficulty conceiving might be related to their body, will cause more psychological distress (in my experience) than it does to women.

REFERENCES

1. Cooper TG, Noonan E, von Eckardstein S, Auger J, Baker HW, Behre HM et al. World Health Organization reference values for human semen characteristics. *Human Reproduction Update*. 2010;16(3):231–45.
2. Stovall DW, Guzick DS, Berga SL, Krasnow JS, Zeleznik AJ. Sperm recovery and survival: Two tests that predict *in vitro* fertilization outcome. *Fertility and Sterility*. 1994;62(6):1244–9.
3. Hebles M, Dorado M, Gallardo M, Gonzalez-Martinez M, Sanchez-Martin P. Seminal quality in the first fraction of ejaculate. *Systems Biology in Reproductive Medicine*. 2015; 61(2):113–6.
4. Mallidis C, Howard EJ, Baker HW. Variation of semen quality in normal men. *International Journal of Andrology*. 1991;14(2):99–107.
5. Reijo R, Alagappan RK, Patrizio P, Page DC. Severe oligozoospermia resulting from deletions of azoospermia factor gene on Y chromosome. *Lancet (London, England)*. 1996;347(9011):1290–3.
6. Sheu G, Revenig LM, Hsiao W. Physiology of Ejaculation. In: Mulhall JP, Hsiao W, editor. *Men's Sexual Health and Fertility*, New York: Springer Science + Business Media; 2014. p. 268.
7. Mann T. *The Biochemistry of Semen*. London: Methuen & Co. Ltd; 1954. p. 258.
8. Huggins C, Scott W, Heinen J. Chemical composition of human semen and of the secretions of the prostate and seminal vesicles. *American Journal of Physiology*. 1942;136(3):467–73.
9. Silber SJ, Ord T, Balmaceda J, Patrizio P, Asch RH. Congenital absence of the vas deferens. The fertilizing capacity of human epididymal sperm. *New England Journal of Medicine*. 1990;323(26):1788–92.
10. Ciliary Dyskinesia, Primary, 1; CILD1 [Internet]. Johns Hopkins University. 2017 [cited 14/12/2018]. Available from: https://www.omim.org/entry/244400?search=Kartagener%E2%80%99s%20syndrome&highlight=kartagener%20syndromic%20s%20syndrome.
11. Davila Garza SA, Patrizio P. Reproductive outcomes in patients with male infertility because of Klinefelter's syndrome, Kartagener's syndrome, round-head sperm, dysplasia fibrous sheath, and 'stump' tail sperm: An updated literature review. *Current Opinion in Obstetrics & Gynecology*. 2013;25(3):229–46.
12. Crosnoe LE, Grober E, Ohl D, Kim ED. Exogenous testosterone: A preventable cause of male infertility. *Translational Andrology and Urology*. 2013;2(2):106–13.
13. Dada R, Gupta NP, Kucheria K. Spermatogenic arrest in men with testicular hyperthermia. *Teratogenesis, Carcinogenesis, and Mutagenesis*. 2003:23;(Suppl 1):235–43.
14. Ford WC, Northstone K, Farrow A. Male employment in some printing trades is associated with prolonged time to conception. *International Journal of Andrology*. 2002;25(5):295–300.
15. De Rosa M, Zarrilli S, Paesano L, Carbone U, Boggia B, Petretta M et al. Traffic pollutants affect fertility in men. *Human Reproduction (Oxford, England)*. 2003;18(5):1055–61.
16. La Vignera S, Condorelli RA, Vicari E, D'Agata R, Calogero AE. Effects of the exposure to mobile phones on male reproduction: A review of the literature. *Journal of Andrology*. 2012;33(3):350–6.
17. Adams JA, Galloway TS, Mondal D, Esteves SC, Mathews F. Effect of mobile telephones on sperm quality: A systematic review and meta-analysis. *Environment International*. 2014;70:106–12.

18. Avendano C, Mata A, Sanchez Sarmiento CA, Doncel GF. Use of laptop computers connected to internet through Wi-Fi decreases human sperm motility and increases sperm DNA fragmentation. *Fertility and Sterility*. 2012;97(1):39–45.e2.
19. Williams AC, Ford WC. Relationship between reactive oxygen species production and lipid peroxidation in human sperm suspensions and their association with sperm function. *Fertility and Sterility*. 2005;83(4):929–36.
20. Ko EY, Sabanegh ES, Jr. The role of over-the-counter supplements for the treatment of male infertility–fact or fiction? *Journal of Andrology*. 2012;33(3):292–308.
21. Keskes-Ammar L, Feki-Chakroun N, Rebai T, Sahnoun Z, Ghozzi H, Hammami S et al. Sperm oxidative stress and the effect of an oral vitamin E and selenium supplement on semen quality in infertile men. *Archives of Andrology*. 2003;49(2):83–94.
22. Benedetti S, Tagliamonte MC, Catalani S, Primiterra M, Canestrari F, De Stefani S et al. Differences in blood and semen oxidative status in fertile and infertile men, and their relationship with sperm quality. *Reproductive Biomedicine Online*. 2012;25(3):300–6.
23. Steiner AK, Hansen MP, Diamond C, Coutifaris M, Cedars R, Legro R et al., editors. *Antioxidants in the Treatment of Male Factor Infertility: Results from the Double Blind, Multi-Center, Randomized Controlled Males, Antioxidants, and Infertility (MOXI) Trial*. ESHRE; 2018; Barcelona: OUP.
24. Jensen TK, Hjollund NH, Henriksen TB, Scheike T, Kolstad H, Giwercman A et al. Does moderate alcohol consumption affect fertility? Follow up study among couples planning first pregnancy. *BMJ (Clinical Research Ed)*. 1998;317(7157):505–10.
25. Minguez-Alarcon L, Chavarro JE, Gaskins AJ. Caffeine, alcohol, smoking, and reproductive outcomes among couples undergoing assisted reproductive technology treatments. *Fertility and Sterility*. 2018;110(4):587–92.
26. Lopez-del Burgo C, Gea A, de Irala J, Martinez-Gonzalez MA, Chavarro JE, Toledo E. Alcohol and difficulty conceiving in the SUN cohort: A nested case-control study. *Nutrients*. 2015;7(8):6167–78.
27. Campagne DM. Can male fertility be improved prior to assisted reproduction through the control of uncommonly considered factors? *International Journal of Fertility & Sterility*. 2013;6(4):214–23.
28. Sanz C. Gym supplements cut sperm count, expert warns. *The Times*. 8 June 2017.
29. Hackney AC, Lane AR. Exercise and the regulation of endocrine hormones. *Progress in Molecular Biology and Translational Science*. 2015;135:293–311.
30. Sansone A, Sansone M, Vaamonde D, Sgrò P, Salzano C, Romanelli F et al. Sport, doping and male fertility. *Reproductive Biology and Endocrinology: RB&E*. 2018;16(1):114.
31. Baran C, Mitchell GC, Hellstrom WJ. Cycling-related sexual dysfunction in men and women: A review. *Sexual Medicine Reviews*. 2014;2(3-4):93–101.
32. Awad MA, Gaither TW, Murphy GP, Chumnarnsongkhroh T, Metzler I, Sanford T et al. Cycling, and male sexual and urinary function: Results from a large, multinational, cross-sectional study. *Journal of Urology*. 2018;199(3):798–804.
33. National Collaborating Centre for Womens and Children's Health. *National Institute for Health and Clinical Excellence: Guidance. Fertility: Assessment and Treatment for People with Fertility Problems*. London: Royal College of Obstetricians & Gynaecologists. National Collaborating Centre for Women's and Children's Health; 2013.
34. European Metrodin HP Study Group. Efficacy and safety of highly purified urinary follicle-stimulating hormone with human chorionic gonadotropin for treating men with isolated hypogonadotropic hypogonadism. *Fertility and Sterility*. 1998;70(2):256–62.
35. Morris G, Cahill D. Gonadotrophin-induced spermatogenesis: Eight years' experience. *Human Fertility*. 2018;21(1):52–75.
36. Depenbusch M, von Eckardstein S, Simoni M, Nieschlag E. Maintenance of spermatogenesis in hypogonadotropic hypogonadal men with human chorionic gonadotropin alone. *European Journal of Endocrinology*. 2002;147(5):617–24.

37. Hull MG, Savage PE, Bromham DR. Prognostic value of the postcoital test: Prospective study based on time-specific conception rates. *British Journal of Obstetrics and Gynaecology.* 1982;89(4):299–305.
38. Glazener CM, Ford WC, Hull MG. The prognostic power of the post-coital test for natural conception depends on duration of infertility. *Human Reproduction (Oxford, England).* 2000;15(9):1953–7.
39. Oei SG, Helmerhorst FM, Bloemenkamp KW, Hollants FA, Meerpoel DE, Keirse MJ. Effectiveness of the postcoital test: Randomised controlled trial. *BMJ (Clinical Research Ed).* 1998;317(7157):502–5.
40. Sadeghi-Nejad H, Farrokhi F. Genetics of azoospermia: Current knowledge, clinical implications, and future directions. Part II: Y chromosome microdeletions. *Urology Journal.* 2007;4(4):192–206.
41. Page DC, Silber S, Brown LG. Men with infertility caused by AZFc deletion can produce sons by intracytoplasmic sperm injection, but are likely to transmit the deletion and infertility. *Human Reproduction (Oxford, England).* 1999;14(7):1722–6.
42. Stouffs K, Vloeberghs V, Gheldof A, Tournaye H, Seneca S. Are AZFb deletions always incompatible with sperm production? *Andrology.* 2017;5(4):691–4.
43. Zhang YS, Dai RL, Wang RX, Zhang HG, Chen S, Liu RZ. Analysis of Y chromosome microdeletion in 1738 infertile men from northeastern China. *Urology.* 2013;82(3):584–8.
44. Talarczyk-Desole J, Berger A, Taszarek-Hauke G, Hauke J, Pawelczyk L, Jedrzejczak P. Manual vs. computer-assisted sperm analysis: Can CASA replace manual assessment of human semen in clinical practice? *Ginekologia Polska (Poznan, Poland).* 2017;88(2):56–60.
45. Lewis SE, John Aitken R, Conner SJ, Iuliis GD, Evenson DP, Henkel R et al. The impact of sperm DNA damage in assisted conception and beyond: Recent advances in diagnosis and treatment. *Reproductive Biomedicine Online.* 2013;27(4):325–37.
46. Alexander NJ, Anderson DJ. Immunology of semen. *Fertility and Sterility.* 1987;47(2):192–205.
47. Walker WH, Cheng J. FSH and testosterone signaling in Sertoli cells. *Reproduction (Cambridge, England).* 2005;130(1):15–28.

Helping women to ovulate 5

OVERVIEW

This chapter covers a systematic approach to women whose periods are absent or infrequent, the WHO Types of ovulation disorders, treatment of ovulatory disorders, and the effectiveness of those treatments.

In Chapter 2, I outlined some of the clinical factors that have a negative impact on ovulation, age, weight and polycystic ovary syndrome being the principle ones. Chapter 3 covers the relevant investigations – measuring prolactin, thyroid-stimulating hormone, timed FSH and LH, progesterone and other hormones such as AMH. In this chapter, I give an account of disorders of ovulation, with treatments and their effectiveness.

5.1 WOMEN WHOSE PERIODS ARE ABSENT OR VERY INFREQUENT

A systematic approach to assessing the problem of amenorrhoea or oligomenorrhoea divides the body into compartments. This compartmentalised approach is described by Speroff, Case and Glass in their comprehensive textbook, *Clinical Gynaecological Endocrinology and Infertility* (1), which I have used since I was sitting my postgraduate examinations. The arcuate nucleus in the hypothalamus, the anterior pituitary and the ovaries are viewed as endocrine sources (the hypothalamic-pituitary-ovarian [HPO] axis), and the endometrium as the end organ (Figure 5.1).

5.1.1 The uterus and vagina

The key functional areas here are the endometrium, the cervix and the vagina. If the history is one of periods which have stopped, secondary amenorrhoea, the likely reasons are infective or traumatic damage to the endometrium. Reviewing the history may assist – past infection with tuberculosis (TB), retained placenta products after miscarriage or overly vigorous curettage after miscarriage are possible causes to consider.

In the complete absence of periods since puberty, primary amenorrhoea, the first steps in investigation are to give the woman a 2-week course of a progestogen drug such as norethisterone or medroxyprogesterone acetate (I prefer the latter – the side effects are less androgenic). The woman should experience vaginal bleeding after stopping the tablets. If she does, it means that the lining of the uterus, the endometrium, is being exposed to oestrogen and that the ovaries are working and the endometrium is normal. It also means that there are no anatomical barriers such as cervical stenosis or a Mullerian anomaly such as imperforate hymen or a more extensive occlusion or developmental failure of the vagina.

If, however, bleeding does not occur, the next step is to give a 4-week course of an oral contraceptive steroid (OCS) which contains oestrogen followed by progestogen. This should produce vaginal bleeding. If she fails to bleed after challenge with an OCS, it points to the anatomical causes listed above, or to a damaged, malfunctioning endometrium. An MRI scan of the pelvis, looking at vagina and uterus will identify anatomical abnormalities. Anatomical causes are beyond the management skill set of most specialists in reproductive medicine, and the young woman should be referred to an appropriate specialist centre for further management.

5.1.2 The ovaries

If menstruation occurs after the OCS challenge, this indicates that the endometrium, when stimulated by oestrogen, is healthy and capable of growth, so that when exposed to progestogen, it bleeds. The cause for absent periods is therefore due to a lack of oestrogen production by the ovaries, either

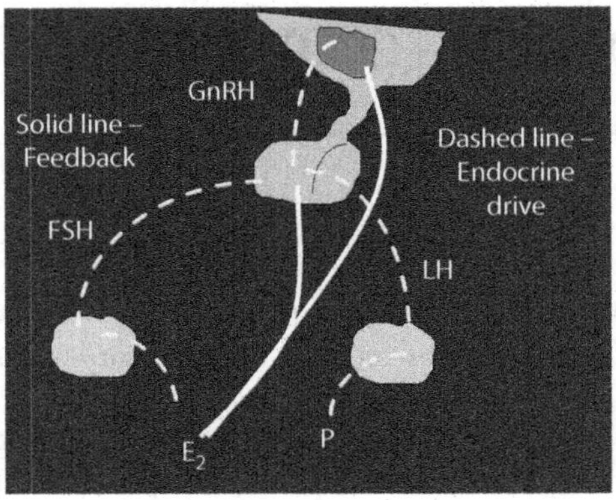

Figure 5.1 The hypothalamic-pituitary-ovarian axis (dashed and solid lines indicating drive and feedback, respectively, respectively).

as a result of ovarian exhaustion of follicles or lack of stimulation by the pituitary. Clarification is provided by measuring levels of FSH.

Ovarian failure is diagnosed if the levels of FSH are extremely high (above 20 IU/L). Raised FSH means that there is drive from the pituitary, but inadequate response from the ovaries. Raised FSH levels are usually accompanied by raised LH levels, in a similar range to FSH, and to low oestrogen levels, usually less than 100 pmol/L. The most common cause of ovarian failure is exhaustion of all responsive follicles in the ovary. Exhaustion of all the follicles usually implies the end of reproductive life, whether this is at the normal menopausal time of 49–51 years or earlier. A premature menopause is usually considered to be before the age of 40, and the causes may be due to chemotherapy or radiotherapy (see Section 9.3 in Chapter 9, for broader coverage of this) or as a result of rare chromosomal/genetic conditions, such as the Fragile X premutation (2). A transvaginal scan of the ovaries will show an absence or depletion of follicles in the ovary, usually less than 6 follicles in each ovary.

Polycystic ovary syndrome. Another ovarian cause of absent periods is polycystic ovary syndrome – a complex syndrome with multiple expressions, the most common being irregular periods, excess weight or obesity, hirsutism and characteristic endocrine parameters. Generally, FSH is normal, LH is elevated (above 10 IU/L and with a 2:1 or 3:1 ratio to FSH), oestrogen levels are normal or moderately raised and androgen levels (testosterone and androstenedione) are moderately elevated in conjunction with lower sex hormone binding globulin levels. In addition, there is insulin resistance, with raised insulin and insulin-like growth factor binding protein levels. The clinical finding which gives the condition its name is the appearance of multiple small follicles scattered around the periphery of the ovary, the polycystic ovary. Polycystic ovaries are not necessary for the diagnosis of PCOS, and a combination of other clinical findings is sufficient to make the diagnosis. Conversely, not all women with polycystic ovaries have PCOS. However, the contribution of the cystic ovary, when present, to the syndrome of PCOS is as follows: normally, in a regular cycle, one follicle moves into dominance and all the other follicles regress as gonadotrophin levels fall (Section 5A.1 provides more detail on this). In PCOS, follicles move out of the phase of being nonresponsive to FSH, developing an antrum and granulosa cells with receptors and becoming responsive. The small amount of FSH present is enough to start this process, but it goes no further – because the follicles are not stimulated enough to develop more receptors to FSH, which would augment the process. But each small follicle does produce a small amount of oestrogen (oestradiol), contributing to the overall picture of PCOS being a state of normal (or slightly elevated) oestrogen levels.

I struggle with the 2003 Rotterdam criteria for diagnosis of PCOS, which focuses on hyperandrogenism and cycle irregularities – 'a syndrome of ovarian dysfunction along with the cardinal features of hyperandrogenism and polycystic ovary morphology' – but with little emphasis on gonadotrophin concentrations, which, in the United Kingdom, is generally understood to be a key feature of PCOS (3). As an aside, it is interesting that, while men do not experience PCOS, not having ovaries (!), there is hyperandrogenism in the male siblings of affected women, expressed as hair loss and premature male pattern balding, occurring under 30 years of age.

Most major reviews of PCOS include a phrase like 'there is generally poor understanding of its aetiology' (4). There is no perfect model to explain all the complexities of endocrine abnormalities seen in PCOS. One model, proposed by Pasquali and developed more fully by others (5,6), addresses many of the endocrine disorders which are found in the syndrome. Pasquali has followed this up more recently, emphasising the impact of insulin resistance and androgen excess, along with associated hyperinsulinaemia and abdominal-visceral obesity (7). This model is graphically illustrated in Figure 5.2a and b. It is postulated that a fundamental pathological process in PCOS is the upregulation of an ovarian cytochrome enzyme, P450c17α, probably as a result of hyperinsulinaemia. The effect of this is to increase androgen levels in the follicular theca cells. These androgens should normally transfer across the follicle wall into the granulosa cells, where under the influence of aromatase enzyme, they are broken down into oestradiol. However, aromatase activity is rate limited; it does not increase in activity no matter how much substrate exists for it. As a result, androgen levels in the ovary increase and spill out into the circulation. In the liver, high androgen levels act perversely to reduce the amount of sex hormone–binding globulin, allowing more free androgen to circulate and affect hair and skin cells. Androgens increase the stimulus for skin and hair cells to grow, producing the unwanted side effects of hirsutism, greasy skin and acne. Androgen feedback to the pituitary gives a continuous steady stimulus to LH, which in turn suppresses pulsatility of GnRH and FSH. However, because there are multiple small follicles in each ovary (see earlier), each producing small amounts of oestrogen, the cumulative overall effect (as far as the body is concerned) on oestrogen production is to keep it normal or even slightly above

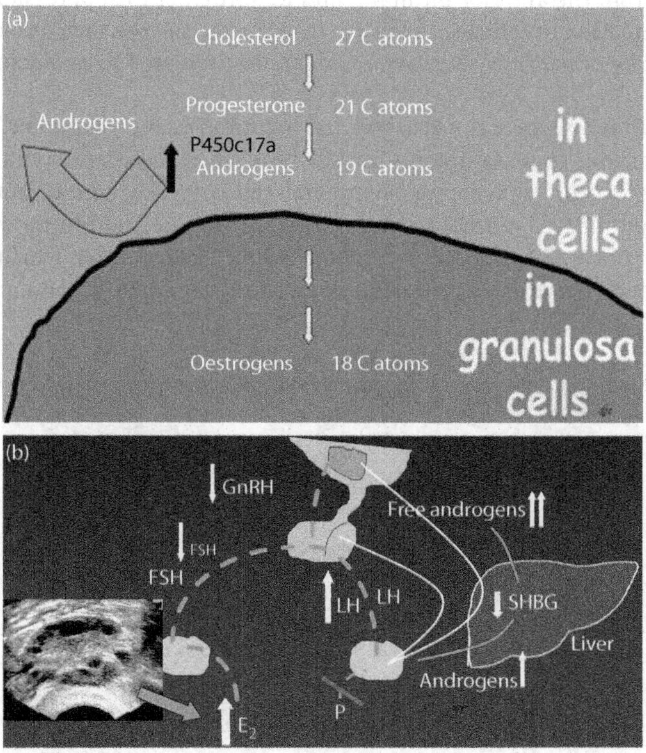

Figure 5.2 (a) Cytochrome P450c17α driving androgen increase; (b) peripheral effects of androgen increase.

normal. As a result, women with PCOS are found to be in a normal oestrogenised state, in contrast to women with hyperprolactinaemia and hypothyroidism, who exist in a low oestrogenised state.

It is important to point out that there are other models to explain the origins of PCOS: these include a neuroendocrine model described by Dumesic (4) and one which uses animal models describing how androgen exposure during fetal life leads to an increased risk of PCOS later in adulthood (8).

Hypogonadotrophic hypogonadism. Finally, if the FSH and LH levels are low (both <4 IU/L) and serum oestrogen levels are low (<100 pmol/L), this points towards a lack of stimulation from the pituitary. There are two main causes – inadequate stimulation from the hypothalamus, or something in the pituitary preventing FSH and LH being secreted (generally a nonsecreting space-occupying mass or a hormonally active tumour). Investigations include measuring other pituitary hormones (prolactin and TSH) which if disordered have a detrimental effect on gonadotrophin hormones and undertaking imaging of the pituitary. Currently, MRI is the favoured imaging approach. Normal investigations results indicate a hypothalamic cause.

Raised prolactin levels. A woman with hyperprolactinaemia classically presents with absence of periods, milk leakage, some loss of axillary and pubic hair, and shrinkage of the gonads, while men present with breast enlargement and erectile dysfunction (Figure 5.3). Remember that some medications (antipsychotics, H2-blockers) can induce hyperprolactinaemia as a side effect – do not forget to ask about these (9). Raised prolactin levels also result from a hormonally active prolactinoma. Prolactinomas occur in 50/100,000 of the general population and are more common in women of childbearing age (10). In this condition, serum prolactin levels usually exceed 2000 IU/L. Figure 5.9a,b illustrates homeostasis and the mechanisms of pathology at work. Within the anterior pituitary, a nidus of prolactin-producing cells begin to produce prolactin in excess. This suppresses GnRH secretion in the hypothalamus, leading to a suppression of FSH and LH secretion from the anterior pituitary, and then further to suppression of oestradiol due to lack of FSH stimulation of follicles. A low oestrogen state then ensues. Prolactin production is normally inhibited by dopamine, secretion of which is stimulated by prolactin in a classic negative feedback loop.

Nonsecreting pituitary tumours can also give rise to problems and symptoms, largely due to their bulk and increase in size. These symptoms include headaches, nausea or vomiting, with occlusion of vision, giving rise to tunnel vision or bitemporal hemianopia. These tumours affect hormone production by the pituitary if normal pituitary tissue is squeezed within the restricted enclosure of the bony pituitary fossa. Bitemporal hemianopia is a serious condition which, if not treated, can lead to permanent optic nerve damage and blindness. I treated a male accountant who presented in

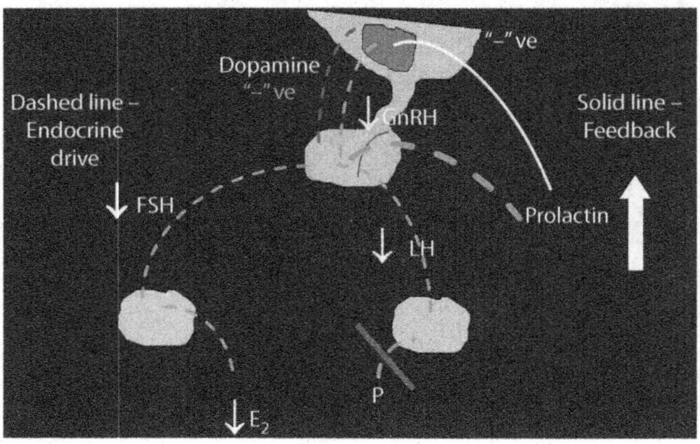

Figure 5.3 How raised prolactin levels suppress gonadal function.

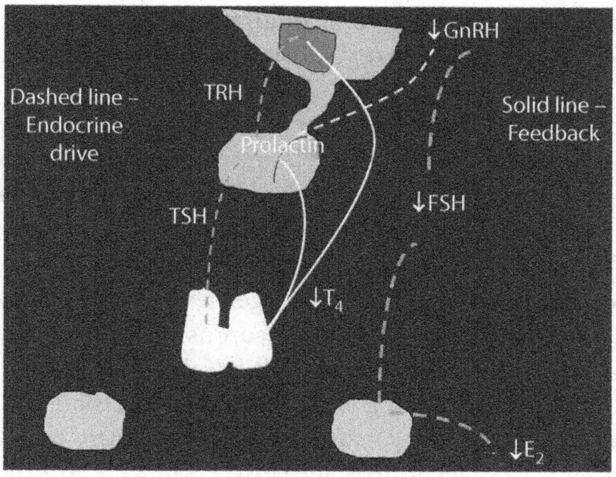

Figure 5.4 How hypothyroidism suppresses gonadal function.

this way – his first symptom was that he could not see the figures in spreadsheets at the outside of the pages he was working on!

Raised thyroid-stimulating hormone (TSH). Another pituitary cause of absent or very infrequent periods is hypothyroidism. Hypothyroidism describes inadequate thyroid gland function, generally due to autoimmune disease or inadequate dietary iodine, and rarely to radiotherapy exposure (Figure 5.4). Low triiodothyronine (T3) and thyroxine (T4) levels cause an increase in thyroid-stimulating hormone and thyrotropin-releasing hormone to drive thyroxine production harder. It was noted earlier that prolactin production is managed by the suppressant effect of dopamine. However, TRH can, in sufficient amounts, exert a stimulatory effect on prolactin secretion, causing hyperprolactinaemia and the suppressant effect on GnRH described above. This is expanded and explained further in Section 5.6.2.

In the appendix to this chapter, details of ovulation are described in greater detail.

5.1.3 The hypothalamic-pituitary-ovarian axis

GnRH secretion occurs in a pulsatile manner. In the first half of the menstrual cycle, this occurs about every 90 minutes. When GnRH is secreted, it passes down neural axons to the hypophyseal portal system and is carried directly to the anterior pituitary. The pulsatility exhibited by GnRH produces pulsatile activity in the gonadotroph cells in the anterior pituitary, producing FSH and LH release in a pulsatile manner. This pulsatility is easily seen in LH levels because of the short half-life of LH, but less so in FSH levels because of its longer half-life, leading to a blunting of any FSH pulses. The benefit of this is that follicles in the ovary are exposed to relatively steady levels of FSH, allowing synchronous development of the granulosa cells and steady exposure of oestradiol to the maturing oocyte. Low LH levels are desirable during this phase, so as not to expose the follicle to unwanted luteinisation. When oestradiol levels are sufficiently high, hopefully having produced a mature, fully grown oocyte capable of fertilisation, the pulsatility of GnRH changes, becoming faster, every 60 minutes. This increased pulsatility specifically stimulates LH-producing gonadotroph cells in the anterior pituitary, and the increased level of LH triggers ovulation. This is a wondrous control system!

5.1.4 The hypothalamus

Hypothalamic causes of failure of menstruation are not common. Conditions include: congenital primary causes such as idiopathic hypogonadotropic hypogonadism – failure of secretion of GnRH and subsequent failure of gonadotrophin stimulation – and an ever rarer variation of this, Kallman syndrome, which includes idiopathic hypogonadotropic hypogonadism and anosmia (a lack of a sense of smell) thought to be due to failure of migration, during early fetal life, of the cells of the arcuate nucleus and the olfactory sensory neurones from their origin in the hindbrain to the hypothalamus.

Other causes are secondary to other pathology – a brain tumour, treated by surgery or radiotherapy, for example. There should usually be a relevant history to point towards the diagnosis.

You will remember reading in Chapter 1 how the HPO axis is the major controller of gonadotrophin function, with neuronal control though kisspeptins from the anterolateral periventricular nucleus. The impact of stress, athletic exercise and disorders such as anorexia nervosa can cause a suppression of the HPO axis, resulting in low oestrogen levels, with infrequent or absent periods. A possible explanation for this lies with the KNDy neurons. These neurons are sensitive to dietary restriction, and (in animal models, at least) this leads to inhibition of *Kiss1* mRNA expression. In animals, dietary restriction thus leads to gonadotrophin suppression and anovulation. In other animal models (rats and chickens), KNDy neurons have been shown to have glucocorticoid receptors, giving a mechanism for stress agents to reduce kisspeptin secretion and cause GnRH suppression (11).

5.2 WHAT ARE THE CATEGORIES OF OVULATION DISORDERS?

The WHO divides the disorders of ovulation into:

- I – Hypogonadotrophic hypogonadism (low gonadotrophins and low oestrogen) – 10%
- II – Gonadotrophin disorder with normal oestrogen (predominantly PCOS) – 85%
- III – Premature ovarian failure (high gonadotrophins and low oestrogen) – 4%–5%

Once other causes of ovulatory dysfunction are ruled out (mostly hyperprolactinaemia and hypothyroidism) (12), women can be allocated relatively easily into one of these three categories.

Anyone can choose how to memorise the pattern of results which determines the category of ovulation disorder present, but memory is not necessary if logic is applied to a knowledge of the HPO axis. So, low FSH and LH levels (generally <5 IU/L and usually <1 IU/L) and low oestrogen indicate an absence of stimulation by the pituitary (WHO Type I). When low levels of oestrogen are found in the circulation, accompanied by raised FSH and LH, this implies that the ovaries are not functioning and probably failing (WHO Type III). Elevated LH, normal levels of FSH and normal levels of oestrogen in serum are consistent with WHO Type II. Low gonadotrophins and high oestrogen is an unlikely finding, but it might be observed in the days coming up to ovulation before the LH surge (dark arrow A in Figure 5.5).

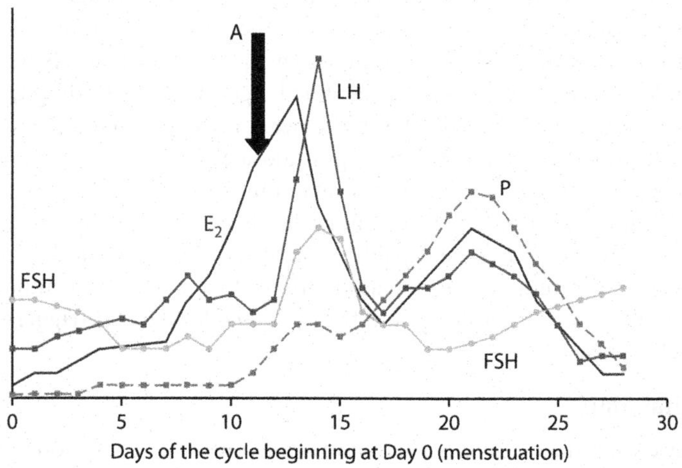

Figure 5.5 Ovarian endocrinology during the menstrual cycle.

5.3 WHAT ARE THE APPROACHES TO TREATMENT?

5.3.1 Medical treatments for disorders of ovulation

The best approach is to consider simple treatments first, and then move on to more complex approaches if these don't work or aren't tolerated. Ovulation induction is the term used to indicate the process of preparation for the development of follicles in cycles where conception will be by sexual intercourse or IUI. The aim of treatment is to recruit one mature competent follicle (mono-folliculogenesis). This contrasts with the process of controlled ovarian superovulation, which is designed to produce multiple follicles (and eggs) for collection as part of assisted conception techniques.

Clomifene citrate is a (relatively ineffective) nonsteroidal anti-oestrogen medication. It works by simulating a low oestrogen status to the hypothalamus, which responds by increasing GnRH pulses to the pituitary, driving FSH production upwards and thereby driving follicular development and ovulation. It only works in the presence of some circulating oestrogen – it is not effective in either WHO Type I or Type III, only in Type II, which includes women with PCOS. Clomifene has both oestrogenic and anti-oestrogenic effects. Centrally, it induces follicular development, but peripherally, in the cervix and in the endometrium, it appears to have largely anti-oestrogen effects, reducing the amount of cervical mucus and the thickness of the endometrium. Clomifene is produced in two isomers, enclomifene [(*E*)-clomifene] and zuclomifene [(*Z*)-clomifene]. Clomifene's half-life is reported as 5–7 days, but the zuclomifene isomer can be present for up to 6–8 weeks afterwards, and certainly beyond the time in which an embryo would be exposed it, should the woman get pregnant. There are, however, no reports of an increase in miscarriage rates or congenital anomalies associated with its use. Clomifene does have side effects which include multiple pregnancy and ovarian hyperstimulation syndrome (OHSS) (see Section 8.7.1 in Chapter 8).

The usual starting dose of clomifene is 50 mg/day from Days 2–6 of the menstrual cycle, 5 days in total. To assess its effectiveness, do a serum progesterone test on Day 21 to confirm ovulation. If serum progesterone levels are low (less than 25 nmol/L), increase the clomifene dose to 100 mg/day in the next cycle and check on Day 21 to see if progesterone levels have responded. If ovulation is not observed on a dose of 100 mg/day, treatment should be discontinued, and the patient should be considered clomifene resistant and moved to a more effective approach. I would not increase the dose of clomifene above 100 mg/day for two reasons:

- Multiple pregnancies then become more likely.
- There is a theoretical risk of an anti-oestrogen effect in the cervix, adversely affecting cervical mucus. Whatever benefit seen in ovulation might be offset by this adverse effect.

It is recommended practice in the 2013 NICE guidelines that clinics should 'offer ultrasound monitoring during at least the first cycle of treatment, to ensure that [the patient] is taking a dose that minimises the risk of multiple pregnancy' (13). An ultrasound scan is done on Day 10 of the cycle. If multiple follicles are seen, the couple is advised not to have sex during that cycle. Multiple pregnancies do occur with clomifene use, but in 28 years of practice, I have never encountered this when using the lowest dose of 50 or 100 mg/day. The multiple pregnancy rate with clomifene is less than 10%, with the majority of these being twin pregnancies. Undertaking a scan is resource intensive, and NHS clinics may struggle to provide this standard.

Clomifene, if well tolerated, can be taken for up to 6 months, but for safety reasons it should not be taken any longer than that. In 1994, Rossing examined women with infertility, looking at the relative risk of (borderline and invasive) ovarian cancers and finding an increased relative risk of 2.3 (95% CI 0.5–11.4) in infertile women taking clomifene, compared with untreated infertile women (14). The relative risk was even greater in women taking clomifene for more than 12 months (relative risk 11.1, 95% CI 1.5–82.3). As a consequence, the Committee for Safety in Medicine in the UK advised that women take clomifene for no more than 6 months. These concerns were refuted

10 years later, which provided lower risk ratios for clomifene [0.82 (95% CI 0.4–1.5)], but caution is clearly on the side of using it for as little time as possible (15).

If clomifene doesn't work, what next?

Metformin is a biguanide and is used in type 2 diabetes mellitus to control hyperglycaemia by decreasing gluconeogenesis and increasing peripheral glucose uptake, overall decreasing insulin resistance. There has been considerable interest in its role in PCOS to assist with or actually induce ovulation. It was common practice at one time for any woman with irregular periods to be put on metformin. This no longer happens, and views on its use are evidence based:

- Metformin is not licensed for ovulation induction and, on its own, appears to have little benefit on ovulation (16).
- It has no measurable effect on weight reduction (17).
- It has no beneficial effect on reducing the skin manifestations of PCOS (17).
- It should be avoided when the body is likely to be dehydrated (e.g. fasting pre-operatively) (18).

Metformin is used in conjunction with gonadotrophin therapy as part of assisted conception. Used in this way, a meta-analysis of seven randomised controlled trials found that metformin improved live-birth rates (odds ratio [OR] = 1.94) and pregnancy rates (OR = 2.25), and reduced cycle cancellation rates (OR = 0.41) (19). Disappointingly, the review considered the studies low quality due to incomplete outcome data and several biases and/or confounders. These findings were replicated by a Cochrane review (20) which similarly showed increased clinical pregnancy rates (confirmed live fetus), although live birth rates were not increased. There was a decreased risk of ovarian hyperstimulation syndrome. This suggests that there is overall benefit in using metformin as a supplement to gonadotrophin-stimulated cycles for assisted conception. Whether to maintain metformin into pregnancy once pregnancy is achieved has been vigorously debated. Some advocate stopping it once the pregnancy test is positive because of concerns about fetal abnormalities. Others support continuing treatment through the first trimester to maintain the environment induced by metformin and to reduce the risk of miscarriage. Reassurance about the lack of increase in the rate of congenital anomalies when continued into the first trimester has recently been published, allowing for greater freedom and evidence of safety (21).

Gonadotrophin therapy is used to prepare women for treatment with intercourse only, with intrauterine insemination (addressed in Section 8.3 in Chapter 8), with *in vitro* fertilisation, and with IVF combined with intracytoplasmic sperm injection (Figure 5.7). The gonadotrophins used are FSH and LH. These drugs can come from menopausal women's urine or from using recombinant technology (using Chinese hamster ovary cells). In practical experience, while pharmaceutical companies vie to produce the most effective product, there is not a lot of difference between one individual drug and another, or between a drug derived from urinary sources or from recombinant technology. In my clinic, for the last 20 years, we have used urinary drugs for ovulation induction therapy, with excellent results over time (Table 5.2).

For the actual treatment process with gonadotrophins (Figure 5.6), the first step is for the woman to contact the clinic when her period starts in the cycle in which she wants to have treatment. She attends the clinic that day or the next for a blood test for serum oestradiol. The level of oestradiol predicts if the woman can start the treatment in that cycle – colleagues in Bristol have shown that oestradiol levels above 200 pmol/L are associated with a higher chance of a cancelled cycle and poorer-quality oocytes being produced (22). Providing the oestradiol level is below 200 pmol/L, FSH injections are started using the lowest starting dose (75 IU gonadotrophin, using a highly purified urinary gonadotrophin), and that dose is used daily. If oestradiol levels persist above 200 pmol/L, a short course of high dose progestogens will normally treat this; if not, trying again in a subsequent cycle is probably the safest option. The plan for the use of daily gonadotrophins is for injections of FSH in a dose of 50–75 IU per day, and this usually starts on Day 1 after the onset of menstruation. The response in the first cycle is normally checked at day 8 of the cycle, measuring serum oestradiol only – not with ultrasound. By then the oestradiol level should have increased twofold, indicating

Ovulation Induction (OI) without a GnRHa

SETTING Reproductive Medicine

FOR STAFF All Reproductive Medicine Staff

PATIENTS All patients using the Reproductive Medicine Service

1. Telephone the nursing staff in the Reproductive Medicine Clinic on the first day of your period for your treatment schedule to be arranged.
   Occasionally we may need to delay the start of your treatment if the critical monitoring stage is likely to coincide with a Public Holiday or if we have already booked all the available treatments for that month.

2. The nursing staff will give you a date and time to have a blood test to check your hormone levels. This will show whether you are ready to start the gonadotrophin injections. You will be telephoned with the results of the test and when to start the injections, when to return for the next blood test and when your first scan is due.

3. Your gonadotrophin injections can be given by your husband/partner who can be taught how to give the injections. The treatment schedule is overleaf. Usually the injections will need to be given for about 10-15 days but occasionally for longer. Occasionally it is necessary to alter the dose of the gonadotrophin injections depending on the results of the scans and blood tests.

4. The treatment will be monitored with scans and blood tests until the eggs are ready to be released. The scans and blood tests will need to be done on several occasions, usually 8, 10 and 12 days after starting your gonadotrophin injections. This will be on a Monday, Wednesday and Friday morning. When the scans and blood test show that your eggs are mature, you will be asked to have the injection of hCG to trigger the release of the eggs. This release will be about 38 hours after the hCG injection, and the best time to have intercourse is the day of the hCG trigger and the following day.

5. You need a blood test to monitor ovulation 7 days after hCG injection.

6. Please ring the nursing staff once you know the outcome of the treatment. They will be able to organise a further cycle of treatment, a follow up clinic appointment or a pregnancy scan as appropriate.

Figure 5.6 Summary of gonadotrophin treatment plan (in nondownregulated cycle – no GnRHa).

some follicular recruitment and development, with oestradiol production. If the expected response does not occur, the dose of gonadotrophin can be doubled and the patient reviewed 5 days later. Changes in oestradiol levels will not be evident in a shorter time span (23). The second and further reviews will include at least two measures of follicular function, a serum oestradiol and a transvaginal scan of ovarian development, measuring numbers of follicles and the sizes of the leading ones.

Table 5.1 Serum oestradiol concentrations by follicle diameter in 73 unstimulated ovarian cycles in which an oocyte was collected, fertilised and transferred.

Follicle diameter (mm)	Number of observations serum oestradiol (pmol/L)	~95% range	Geometric mean
10	5	128–587	274
11	6	148–677	316
12	13	170–781	365
13	21	196–902	421
14	26	227–1040	486
15	33	262–1201	561
16	34	302–1385	647
17	36	348–1599	746
18	23	402–1845	861
19	26	463–2129	994
20	22	535–2456	1147
21	19	618–2834	1323
22	12	713–3271	1527
23	2	822–3774	1762
24	6	949–4355	2033

Source: Based on data from Cahill DJ et al. *Human Reproduction (Oxford, England)*. 2000;15(9):1909–12.

Rather than using guesswork to combine the information from these two sets of results, we use data derived from unstimulated 'natural' cycles with competent oocytes (fertilised and transferred as embryos) to help clinic staff interpret the results and plan the next stages of treatment. These data are presented in Table 5.1 (24). When it is estimated, using these measurements, that a mature oocyte is likely to be ovulated, ovulation can be induced by LH (but hCG is used in practice as it is more readily available, biochemically works just like LH and is almost chemically identical).

Over the years, the treatment approach has been very similar to that used in Figure 5.6. Alterations occurred following one year's experience with a success rate of 29%, but there were unexpectedly higher multiple pregnancy rates. When reviewed, these cycles were all found to be in women under the age of 35, with serum oestradiol concentrations in excess of 2000 pmol/L. We changed our approach to cancellation subsequently to recognise this (leading to more cancellations but eliminating higher multiple pregnancy rates), and then, of course, pregnancy rates fell in response.

In hypogonadotrophic hypogonadism, there is no GnRH activity, indicated by low levels of FSH and LH (usually <4 IU/L), and pharmacological methods to mimic it release the woman's own FSH and LH. This can be achieved by using a pulsatile delivery system of GnRH. Such systems (usually portable battery-operated pumps) were available for many years, and good success rates were associated with their use. However, the manufacturers withdrew them for commercial reasons about 10 years ago, something very regrettable, and now the only option is to use exogenous gonadotrophins to stimulate ovulation. These are less physiological and therefore require as much monitoring as any other form of ovulation induction.

The range of disorders treatable by ovulation induction (with or without IUI) is limited to women with healthy tubes. Ideally, all women should have a test of tubal patency before embarking on an ovulation induction agent – but that is not a good use of resources. I tend to allow women with normal or low Chlamydia antibody titres to have clomifene treatment without confirming tubal patency. However, before ovulation induction using gonadotrophins, a much more expensive treatment, I ensure that all with a normal Chlamydia titre have a hysterosalpingogram, unless there is a clinical history of pelvic or abdominal infection or past history of abdominal surgery, in which case, a laparoscopy and tubal insufflation is needed.

1.18 Instructions for Patients about Ovulation Induction (OI) with a GnRH agonist

SETTING	Reproductive Medicine
FOR STAFF	All Reproductive Medicine Staff
PATIENTS	All patients using the Reproductive Medicine Service

1. Telephone the nursing staff in the Reproductive Medicine Clinic on the first day of your period for your treatment schedule to be arranged (phone number). Occasionally we may need to delay the start of your treatment if the critical monitoring stage is likely to coincide with a Public Holiday or if we have already booked all the available treatments for that month.

2. On day 19 of your cycle you will start to take 2 norethisterone tablets a day and continue these for 7 days. A period will start a few days after finishing the tablets.

3. On day 21 of your cycle (2 days after starting your norethisterone tablets) you will start using the GnRHa spray taking one sniff at 8am, 12 midday, 4pm and 8 pm and 2 sniffs last thing at night before you go to bed. The spray will prevent your own hormones from interfering with the treatment.

4. You will have been given an appointment date to have a blood test at least 2 weeks after starting your GnRHa spray. This tests whether you are ready to start the gonadotrophin injections. You will be telephoned with the results of the test and when to start the injections, when to return for the next blood test and when your first scan is due. Continue to use the GnRHa nasal spray while having your gonadotrophin injections.

5. Your husband/partner can be taught to give the injections; alternatively, you could administer the injections yourself. If there are any problems regarding this, please discuss with the nurses. The treatment schedule is overleaf. The injections will need to be given for about 10-15 days but occasionally for longer. It may be necessary to alter the injection dose depending on the results of the scans and blood tests.

6. The treatment will be monitored with scans and blood tests until the eggs are ready to be released. The scans and blood tests will need to be done on several occasions, usually 8, 10 and 12 days after starting your gonadotrophin injections. This will be on Monday, Wednesday and Friday mornings. When your scans and blood tests show that your eggs are mature, you will be asked to have the hCG injection. Your eggs will be released about 38 hours after the hCG injection. The chance of conception is highest if you have intercourse on the day of the hCG trigger and the day after.

7. When you have had the injection of hCG to release your eggs, you should stop using the GnRHa spray. You will need to have a drug given by vaginal pessary every 12 hours starting 2 days after the hCG trigger, until your period comes or your pregnancy test is positive. This is to keep the lining of your uterus stable to support the early stages of any pregnancy.

8. Please ring the nursing staff once you know the outcome of the treatment. They will be able to organise a further cycle of treatment, a follow up clinic appointment or a pregnancy scan as appropriate.

Figure 5.7 Summary of gonadotrophin treatment plan (in cycles downregulated with GnRHa).

Induction of ovulation with gonadotrophins can be achieved with or without a GnRH agonist. A specimen treatment plan for a cycle of treatment not downregulated using a GnRHa is given in Figure 5.6. Using a GnRHa in addition to FSH and LH has the advantage of providing control of the timing of ovulation (as in Figure 5.7). GnRHa suppresses ovarian activity, delaying ovulation, even in the presence of the administered exogenous FSH and LH. An hCG injection is administered when ovulation is needed, producing ovulation with a 36 hour period. Use of GnRHa is indicated when there is a necessity to deliver sperm to the uterus, whether by sexual intercourse or by IUI, close to ovulation. However, this level of exactitude is usually only required for IVF. GnRHa is not required for ovulation induction alone, and reassuringly, evidence shows that introducing sperm up to 36 hours after ovulation does not adversely affect pregnancy rates after sexual intercourse and IUI (25). Practice does vary between clinics, however, and many routinely use GnRHa to influence and schedule the day of ovulation and most publicly funded clinics schedule IUI Monday to Friday; GnRHa is also used if a woman has previously spontaneously ovulated in a treatment cycle and the opportunity was missed to introduce sperm.

While there are positive benefits to the use of GnRHa, it does cause a suppression of ovarian activity, thus requiring more exogenous gonadotrophins than otherwise, and there is tenuous evidence that GnRHa may also independently suppress ovarian gonadotrophin activity (26).

A further development in the use of GnRH analogues is to use GnRH antagonists. GnRH antagonists act by receptor blockage, competitively inhibiting gonadotroph cells and immediately suppressing gonadotrophin secretion (27). They are rapidly reversible, which is advantageous. While GnRH antagonists have been around since GnRH agonists were introduced (28), their use in clinical practice was halted because the early forms provoked very strong histamine responses. Now that those problems have been overcome, the advantages of these drugs are evident. Fewer FSH injections are needed because the introduction of the antagonist in the latter part of the ovarian stimulation makes use of the body's own FSH drive as well as the exogenous injected drug. The combined benefits are a simple short protocol and a reduced incidence of ovarian hyperstimulation syndrome with the ability to start the next treatment cycle immediately (27).

There is also the advantage of speed, which can be critical in the case of women with cancers who need to start ovulation induction immediately to facilitate the start of chemotherapy. There are disadvantages: less flexibility in programming of cycles and a reduction in clinical pregnancy rates, probably largely related to fewer oocytes being collected. We have shown that this reduction is not easily addressed, certainly not just by increasing the gonadotrophin dose (29). A model for GnRH agonist use in illustrated in Figure 5.8a; it is not generally used for intrauterine insemination, only for *in vitro* fertilisation. A model for GnRH antagonist administration, generally for *in vitro* fertilisation, is shown in Figure 5.8b. It shows stimulation with exogenous gonadotrophins from the beginning of the cycle, with the introduction of the GnRH antagonist on day 7 or so. Note that Figure 5.8b uses hCG as the stimulus for triggering ovulation.

5.3.2 Surgical treatments

Laparoscopic ovarian diathermy is the only surgical treatment available to augment ovulation and is only useful in polycystic ovary syndrome. This technique has waned in popularity over recent years. It involves a laparoscopy, and then insertion of a diathermy needle deep into the cortex of the ovary, activating the diathermy for 4 seconds. This is repeated into the ovary 4–6 times. The intended benefit from this procedure is to destroy some or a lot of the androgen-producing tissue in the cortex of the ovary, thereby rectifying the androgen stimulus which is such an important part of the pathology of polycystic ovary syndrome.

5.4 HOW EFFECTIVE ARE THE DIFFERENT TREATMENTS?

Selection of patients is one of the key factors in the success of these treatments. If you only treat women under 35, whose BMIs are under 25 and whose partners have perfect sperm, then you will get very good results. But real life isn't like that, and our patients range up to 40 years of age, with BMIs up to 30 (limits imposed by our local health authority for funded treatment), and men whose sperm counts are often not perfect. There can be considerable difference between different clinics' pregnancy rates, and it is often due to these factors.

Clomifene therapy has poor results in terms of conception – rates of 2%–10% are quoted dependent on whether conception is with intercourse or IUI – higher with IUI (30). Other studies have shown

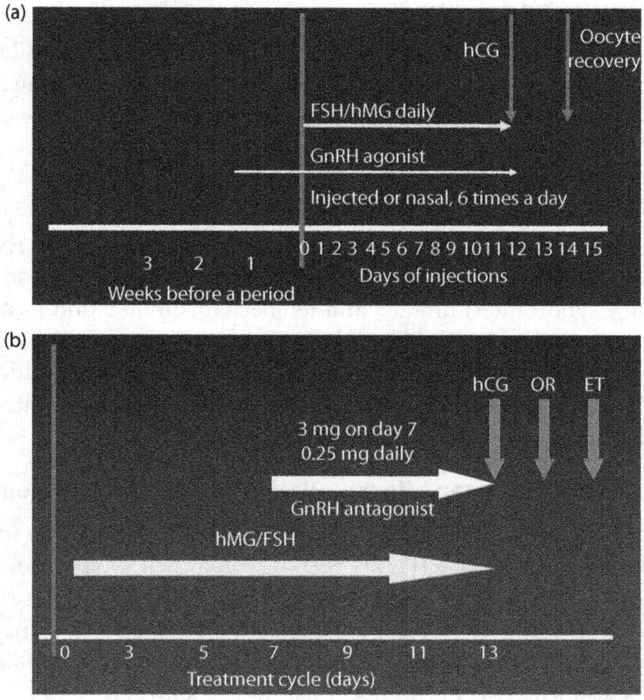

Figure 5.8 (a) Protocol used for GnRH agonist administration; (b) protocol used for GnRH antagonist administration.

some what higher pregnancy rates but also indicated that these were no better than with expectant or 'no active' management (31). Despite its low success, cost and simplicity make it attractive to use.

Ovulation induction therapy with gonadotrophins offers higher pregnancy rates than clomifene citrate, and success rates in our clinic for ovulation induction alone, and with IUI, are presented in Table 5.2. These data stretch back over 4 years, but temporal trends beyond that are broadly consistent with success for ovulation induction being around 15% and ovulation induction and IUI being closer to 20%. National data for women up to the age of 40 are pregnancy rates between 12% and 14% (32).

Surgical treatment (*laparoscopic ovarian diathermy*) does appear to bring about the desired hormonal changes, with falling levels of testosterone and then of LH being observed (33). Return of cycle regularity and reasonable pregnancy rates are observed (34). However, in randomised controlled trials (RCTs), laparoscopic ovarian diathermy has been found to be no better than gonadotrophin use (35). There is a balance, then, to be struck between the risks of laparoscopy and the fact that multiple pregnancy rates are lower with laparoscopic ovarian diathermy use than with gonadotrophins (35).

Table 5.2 Pregnancy rates after ovulation induction with gonadotrophins (Reproductive Medicine Clinic, Bristol).

	2013	2014	2015	2016
Ovulation induction only (sexual intercourse)	14%	25%	8%	38%
(n = number of completed treatment cycles)	25	8	12	13
Ovulation induction and IUI (n = number of	25%	10%	19%	9%
completed treatment cycles)	65	42	26	32

5.5 HOW THEY ALL WORK TOGETHER

Table 5.3 lists all the hormones produced in the anterior and posterior pituitary which influence follicular development, along with GnRH, together with their site of production, secretion and action. There is also some information on the molecular structure, half-life of the molecule and how it is regulated.

5.6 CLINICAL CONDITIONS WHICH CAN AFFECT OVULATION

There are three clinical conditions which can affect and prevent ovulation. Hyperprolactinaemia and hypothyroidism are linked in their mechanisms of action, and both lead to a low oestrogen state. Polycystic ovary syndrome is unique, and its mechanism of action is still not fully understood. In this section, one theory of understanding is proposed, and like many theories, it has deficiencies, but it serves to address much of what is manifested in the condition. In contrast to the other two, PCOS leads to a normal to moderately elevated oestrogen state, and the long-term consequences of this condition are quite different.

5.6.1 Clinical conditions which can affect ovulation: Hyperprolactinaemia

Review again the process of hypothalamic-pituitary-ovarian axis function in the production of oestradiol first and then progesterone (Figure 5.1). A woman with hyperprolactinaemia presents with absence of periods, milk leakage, some loss of axillary and pubic hair and shrinkage of the gonads, while men will present with breast enlargement and erectile dysfunction (and failure of spermatogenesis). Remember that some medications (antipsychotics, H2-blockers) can induce hyperprolactinaemia as a side effect – do not forget to ask about these (9). Raised prolactin levels also result from a hormonally active prolactinoma. Prolactinomas occur in 50/100,000 of the general population and are more common in women of childbearing age (10) and serum prolactin levels are usually 2000 IU/L or more.

Prolactin control as illustrated in Figure 5.9a is maintained by largely negative inhibitory control by dopamine, produced in the hypothalamus which is transferred to the pituitary through the portal system. When a prolactin-secreting tumour, generally a 'microprolactinoma', begins to function or, under the effect of prolactin inducing medications, the direct effect on breast tissue is galactorrhoea. The normal inhibitory effect of dopamine in the pituitary fails, and prolactin proceeds to suppress GnRH secretion either as a pulse reduction in amplitude or, less likely, frequency. This GnRH suppression leads to failure of follicular development, suppression of oestradiol, no endometrial growth, no ovulation and amenorrhoea. Figure 5.9b illustrates this, as well as the downward arrows indicating lower levels of stimulation. Hyperprolactinaemia then is a low-oestrogen state, and women should not be allowed to continue in for prolonged periods, as, it will lead to oestrogen deficiency and osteoporosis, well before the normal age of the menopause (36).

Treatment of raised prolactin due to a prolactinoma is generally medical in the first instance, using bromocriptine, a vinca alkaloid or a newer preparation, cabergoline. Cabergoline causes fewer side effects than bromocriptine (headaches being the most pronounced), but they both cause nausea and vomiting. Treatment will reduce prolactin almost immediately, though effects on the hypothalamic-pituitary-ovarian axis will not be evident for 6–8 weeks. If medical treatment is not successful, then a transsphenoidal hypophysectomy (removal of the pituitary, using a route up through the nasal passages and across the sphenoid sinus) may be required. This is a very destructive procedure and the patient will almost certainly need lifelong replacement of most if not all the pituitary hormones. If hyperprolactinemia is caused by medication prescribed by other medical specialties, ask the patient to seek advice from the prescriber as to alternatives.

5.6.2 Clinical conditions which can affect ovulation: Hypothyroidism

Hypothyroidism is a condition characterised by a generally reduced metabolic state, with clinical features such as somewhat coarser skin, intolerance to cold, diminished energy, loss of the outer half of the eyebrows and general hair thinning and, for some women, anovulation experienced as

Table 5.3 The hypothalamic and pituitary orchestra (with specific reference to ovarian function).

Hormone	Site of production	Site of secretion	Structure	Half-life	Regulation	Site of action	Function
GnRH	Arcuate Nucleus	Nerve ending of GnRH neurone	Decapeptide	2–4 min	GnRH, FSH, LH, E_2	Anterior pituitary	Control of gonadotrophin release
FSH	δ Basophilic	Anterior	Glycoprotein	16 hours	GnRH, E_2	Granulosa cells	E_2 production
LH	Gonadotroph	pituitary	Glycoprotein	3–4 hours	GnRH, P	Theca cells	Progesterone prod.
Oxytocin	Hypothalamus	Posterior	Nonapeptides	4 min	Suckling	Breast, uterus	Uterine, myoepithelial cells
Vasopressin		pituitary	(two a.a.s. differ)	Up to 30 min	Pain, Δ blood volume		Osmolality, blood volume
TSH	β2 Basophilic	Anterior	Glycoprotein	60 min	TRH (tripeptide)	Thyroid adrenal	Corticosteroid
ACTH	Gonadotroph	pituitary	Protein	10 min	CRH (41 aa peptide)	Cortex	production
Prolactin	α Acidophilic	Anterior	200 aa protein	15–20 min	Dopamine−ve, TRH+ve	Breast	Initiation of lactation
GH	Gonadotrophs	Pituitary	Protein	Up to 20 min	GH-RH (44 aa peptide) Somatostatin	Protein synthesis Fat store mobilisation	Generally required in patients with IHH for folliculogenesis

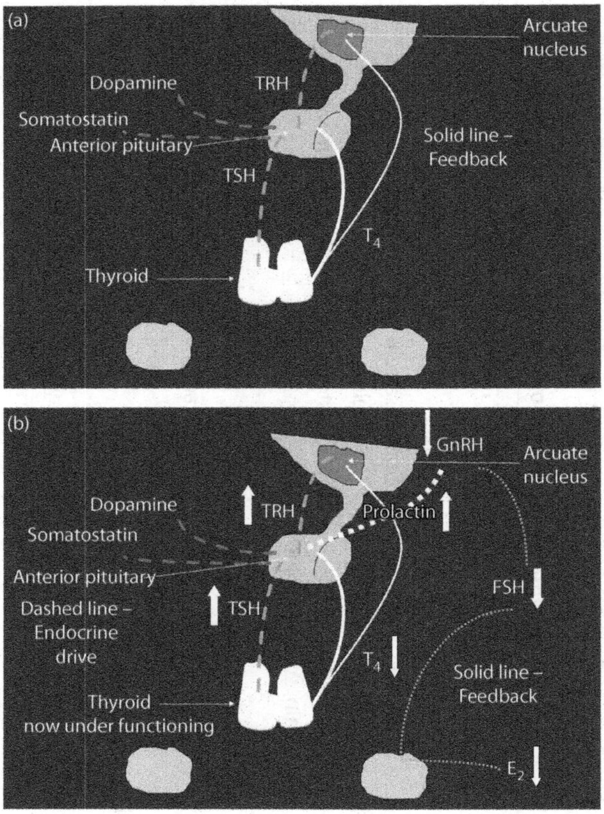

Figure 5.9 (a) Normal thyroid function in homeostasis; (b) abnormal thyroid function: hypothyroidism.

amenorrhoea. Hypothyroidism can be caused by surgery following a partial thyroidectomy, it may result from an autoimmune reaction to the thyroid cells, rarely from radiotherapy exposure, and worldwide, it occurs in states of iodine deficiency, particularly in locations further from the sea, as fish is a good source of iodine (when I grew up in Ireland, I can recall people in the Midlands using iodised salt to prevent iodine deficiency).

Thyroid hormone production is controlled by a feedback loop from the thyroid gland to the pituitary to the hypothalamus and back to the thyroid gland again. High levels of thyroid hormones T4 and T3 (thyroxine and tri-iodothyronine) suppress release of TRH (thyrotrophin-releasing hormone) in the hypothalamus, which in turn reduces release of TSH (thyroid-stimulating hormone) in the pituitary. Without the stimulating effect of TSH on the thyroid gland, production of T4 and T3 falls. Low T4 and T3 levels provoke the hypothalamus to release TRH, and so the circuit continues to maintain the ideal homeostatic state. See Figure 5.10a for illustration.

When T4 production begins to fall (for any of the reasons above), the system tries to correct this by increasing TRH and TSH drive (which is why we measure TSH when we are looking for hypothyroidism; it is much easier to measure than very low levels of thyroxine). The effects of the increase in TRH has an unexpected and inexplicable effect as it also has a positive and stimulatory effect on prolactin. This in turn brings about a suppression of GnRH pulsatility and of FSH production and ultimately of oestradiol production in the ovary, leading to suppression of ovulation and amenorrhoea in a low-oestrogen state, like hyperprolactinaemia (see Figure 5.10b).

Treatment of hypothyroidism is relatively straightforward. Start thyroxine (usually 50 μg daily) immediately, and measure TSH levels within a month. They should be normal - and if not, increase the dose by 25–50 μg. Once TSH levels are normal, cycles should return within 4–6 weeks. Remember

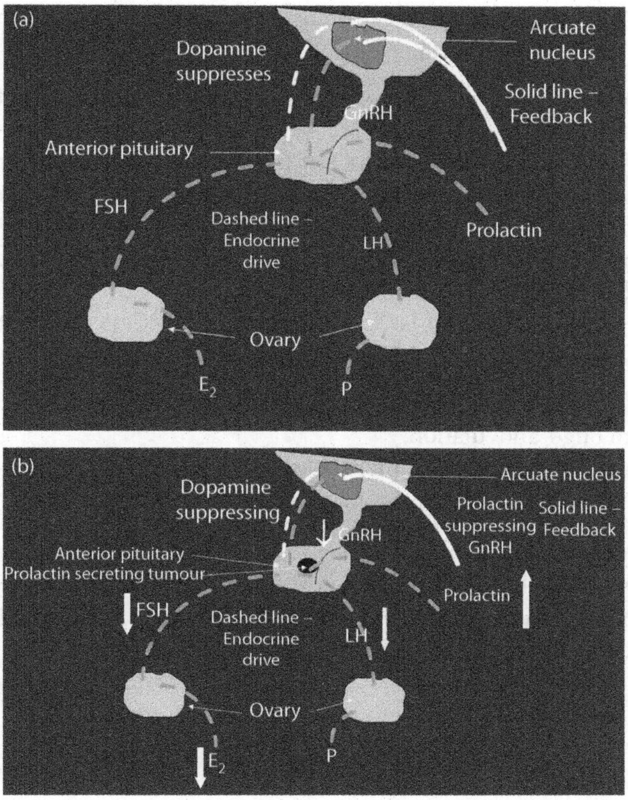

Figure 5.10 (a) The hypothalamic pituitary ovarian axis – prolactin homeostasis; (b) the hypothalamic pituitary ovarian axis – hyperprolactinaemia and a microprolactinoma.

that if the woman conceives on thyroxine, she will almost certainly require an increase in dose within the first few weeks – so she should be directed to inform her prescribing doctor immediately on confirming that she is pregnant, recognising she might otherwise not interact with health care providers until 8–10 weeks amenorrhoea.

5.6.3 Clinical conditions which can affect ovulation: Polycystic ovary syndrome

Polycystic ovary syndrome (P-C-O-S, not PeeCOS, a pronunciation I detest) presents in several ways, and in both thin and obese women. There is the classic description of a woman who (amongst other signs and symptoms) is overweight and has greasy skin, with some or excessive facial (and other) hair, including a male-pattern abdominal hair, with infrequent or absent periods. Women with PCOS are generally found to be in a normal oestrogenised state, in contrast to women with hyperprolactinaemia and hypothyroidism, who exist in a low oestrogenised state.

However, the truth is that the clinical manifestation is very varied, with women of all shapes and sizes presenting. In fact, infertility in women who have PCOS and who are thin are the most difficult to treat to get pregnant, as their response to gonadotrophin therapy is often difficult to predict and they often for over-respond easily in the form of multiple follicular development.

The variety of clinical presentations has necessitated the formation of diagnostic criteria for use by the clinician. The Rotterdam 2003 criteria (3), already referred to above, call for two of these three features:

1. Oligo- or anovulation
2. Clinical and/or biochemical signs of hyperandrogenism

3. Polycystic ovaries and exclusion of other etiologies (congenital adrenal hyperplasia, androgen-secreting tumors, Cushing's syndrome)

Importantly, to fulfil these criteria, it is not necessary to have an ultrasound finding of polycystic ovaries.

Despite all the efforts in research over the last few decades, there is still no clarity about the understanding of PCOS – I love this sentence: 'there is generally poor understanding of its etiology' (4). The model I find most satisfying and useful is one that centres on the ovary, its production of excess androgens and the endocrine and physical effects of that (5,6). That model is described fully in Chpater 5.1.2 and the reader is referred there.

Ricardo Azziz suggests that several (at least four) types of PCOS manifestation may occur in women (37):

1. 'hyperandrogenism oligo-anovulation polycystic ovarian morphology;
2. hyperandrogenism oligo-anovulation;
3. hyperandrogenism polycystic ovarian morphology;
4. oligo-anovulation polycystic ovarian morphology'

Treatment of polycystic ovary syndrome varies depending on what the underlying reason for presentation is. It should be borne in mind that there is a disputed link between to endometrial atypia and cancer (37). Given the risk of endometrial neoplasia, it might be wise to induce a withdrawal bleed once every 3 months, using medroxyprogesterone acetate or similar. Women with PCOS also have an increased risk of type 2 diabetes and cardiovascular disease, the management of which falls other specialisms.

5.7 CASE STUDIES

5.7.1 Case study: Polycystic ovary syndrome

Jane is 26. She has been married for 3 years. She used an oral contraceptive pill for contraception from when she was 20 until 2 years ago. She and her husband then decided they wanted to start a family. Since coming off the Pill, her periods have become irregular and unpredictable, and sometimes she has not had a period for 2 to 3 months. She is happily married and has a steady job that she likes, although it is occasionally stressful. Almost 2 years have passed without getting pregnant, so she asked her GP to refer her to see a specialist.

While awaiting an appointment to see a specialist, Jane's GP did some blood tests; the first in the early part of her menstrual cycle and the second about 5–10 days before her next period was due. These showed absence of ovulation so her GP prescribed a mild ovulation-stimulating agent, clomifene citrate. Despite taking this for 3 months, there was no evidence of ovulation on blood tests. When the specialist saw her, she had an ultrasound scan of her ovaries done. This showed the characteristic appearance of polycystic ovaries. The ultrasound results agreed with the blood tests that had already been taken. As Jane was slightly overweight, the specialist suggested that she start a weight-reducing diet; take smaller, regular meals and start treatment with metformin.

She began taking metformin, one tablet a day. After 3 months, this was increased to two tablets a day. She had some bowel upset at this dose and it was decreased back to one tablet a day. After 6 months on this treatment, her cycles were quite regular. At 7 months, her period was late, and a pregnancy text was positive. As she had already been advised to do, she stopped the metformin tablets.

5.7.1.1 Comments, investigations and management outlines

Stressful situations can cause cycle irregularities. The stress needs to be substantial, for example, recent bereavement, moving countries or beginning a new job.

- First blood tests where periods are irregular: early follicular phase follicle-stimulating hormone and luteinising hormone and other tests for thyroid-stimulating hormone and prolactin

- Her results were:
  FSH: 4.5 IU/L (normal) (IU = international unit, a measure of the hormone)
  LH: 15.9 IU/L (raised)
  TSH and prolactin were normal
- *Second tests*: The crucial test done at this time is progesterone: it was 4 nmol/L (reduced). Any result below 25 nmol/L is likely to mean that ovulation hasn't occurred (or perhaps it has, but the response of the luteal cells in [what was] the follicle is not functioning well enough [producing enough progesterone] to support the endometrium).

Combined with the ultrasound result, the diagnosis is most probably polycystic ovarian syndrome. Metformin reduces excess insulin, LH and testosterone concentrations in the blood (39). The GP in this case *did* prescribe clomifene, but many GPs nowadays will not prescribe this, because they do not have access to the monitoring recommended by NICE guidelines (13). Women with PCOS who have irregular or prolonged cycles will often get a return of ovulation and regular periods on clomifene, though caveats exist for its use, as outlined in Section 5.3. It increases the chance of ovulation and may improve the quality of the eggs that are released. At least 20% of women will conceive within the first 6 months. Temporary mild gastrointestinal side effects, such as nausea or diarrhoea, can sometimes occur with increased doses of metformin, and the dose may need to be adjusted/reduced to take account of this.

5.7.2 Case study: Hyperprolactinaemia

Shilpa is 32 and she married Raj 4 years ago after they met at graduate school. She studied mathematics as an undergrad and did business studies in her second degree, and now works in an accountancy firm. Raj did a postgraduate degree in chemistry and now works for an international industrial chemical firm. They are very happy and are interested in having a baby. Over 2 years ago, Shilpa noted her periods first became light, then infrequent, and they stopped completely a year ago. Her undergraduate science background tells her this isn't right, but there is little time to set aside to pursue this, as work is all consuming, and both she and Raj come home exhausted at the end of each day.

Over the last 6 months, she has noted an increase in migraines or tension headaches, which now affect her most days, and she is struggling at work when looking at spreadsheets on her computer screen. She suspects that she needs glasses and makes an appointment for an optician on Saturday. The optician puts her through a range of tests, including visual acuity and fields of vision. At the end of the tests, the optician tells Shilpa that her visual acuity is normal and she doesn't require glasses, but the optician notes an abnormality in her field of vision. The optician tells her this is something she needs to attend to quickly, despite work pressures. The optician provides Shilpa with a report of the findings of the test to take to her own family doctor.

She gets to see the doctor on the following Wednesday, and the doctor takes a history of all the events of the previous 18 months to 2 years. The doctor examines her and confirms the optician's findings. She arranges an urgent appointment with an endocrinologist at the local hospital and an MRI scan of Shilpa's head. The doctor also arranges several blood tests.

When Shilpa sees the endocrinologist 3 weeks later, the MRI and the blood tests results are available. The specialist tells Shilpa that she is producing too much of one particular hormone and needs to start medication to suppress that hormone immediately. A formal visual field test is performed by a technician at the clinic, and Shilpa collects the medication from the pharmacy and starts to take it, doing so once a week. She gets an appointment for 6 weeks' time to be reviewed at the clinic.

The medication caused her some sickness and abdominal pain in the first week, which then went away. She felt no different for the first 3 weeks, but soon after that, she noted the headaches were improving. At the review, the visual fields were noted to be improved, though not completely normal as yet. Further blood tests were taken. She was encouraged to keep on the medication, which she agreed to do. Within 4 weeks (10 weeks after taking the medicine), she had a period.

5.7.2.1 Comments

In this case, the history and physical findings seem to scream a pituitary tumour causing pressure on the optic nerve with gonadotrophin suppression. The absence of breast leakage, galactorrhoea, is a distractor from hyperprolactinaemia, but the headaches and visual disturbances are red-flag symptoms. The MRI showed a small space-occupying lesion, measuring 4 mm in diameter in the anterior pituitary. This examination cannot distinguish between a hormonally active tumour and an inactive one.

The blood tests showed:

FSH	1 IU/L	low
LH	2.5 IU/L	low
Oestradiol	120 pmol/L	low
TSH	2.5 IU/L	normal
Prolactin	2400 mU/L	raised significantly

These results support a diagnosis of a space occupying prolactinoma for which medical intervention with dopamine agonist therapy should be started. The choice is between bromocriptine (daily therapy, recognised to cause side effects, long established) and cabergoline (newer treatment, weekly therapy, fewer side effects). Prolactin levels should fall within 3–4 weeks, though return of menstruation will follow reactivation of the hypothalamic-pituitary-ovarian axis. If prolactin levels fail to respond to a dopamine agonist, then surgery may be indicated after which many patients require long-term thyroid, corticosteroid and oestrogen/progestogen replacement therapy.

5.8 SUMMARY

A cohort of women will present with infrequent or absent periods, or with already identified problems with ovulation. In this chapter, I take a systematic approach to addressing these issues and consider the various body compartments where these problems arise: the arcuate nucleus in the hypothalamus, the anterior pituitary and the ovaries, with the functioning or malfunctioning of the ovaries (disordered ovulation) being expressed in menstrual irregularities. Localised pathology of the uterus, cervix and vagina as it pertains to primary and secondary amenorrhea is also addressed.

The three major conditions causing irregular or absent periods (PCOS, hyperprolactinemia and hypothyroidism) are explored. No aetiological model fully explains the dysfunction noted in PCOS.

I describe the main medical interventions used in reproductive medicine for ovulation disorders: oral anti-oestrogen therapies, biguanides and injectable gonadotrophins. 'Recipes' for treatment approaches are provided as well as success rates achieved in a nonresearch, outpatient setting.

APPENDIX: PHYSIOLOGY OF OVULATION

Ovulation is a complex process which, when examined in detail, provokes disbelief that any of us are here to read about it. There is an interactive physiological process throughout human ovulation which is controlled and modulated by numerous signals. The ovulatory or menstrual cycle is exactly that: it is cyclical and, in a sense, never ending. For our description of the process, we step onto this never-ending cycle at the beginning of menstruation, traditionally the time considered the beginning of each menstrual cycle.

At the onset of menstruation, several dynamic (in the sense of not static) forces are already at play, carried over from the previous cycle:

- Following collapse of the corpus luteum, serum levels of progesterone and oestrogen are falling.
- Oestrogen support to the endometrium is falling; vessels at the base of the endometrium are going into spasm, leading to tissue death and loss of connective tissue support.

- As oestrogen levels falls, the pituitary, responding to the feedback loop, begins to secrete FSH. Even before menstruation begins, this leads to recruitment of follicles which are 3–4 mm in size and which have already developed FSH receptors.
- In the endometrium, the spasm in the basal endometrial vessels is released and arterial blood rushes in, disrupting the tissue above it and causing it to shear off, resulting in menstruation.
- The tiny arterial blood vessels at the base of the endometrium continue to bleed until oestrogen levels rise, consequent on FSH drive to the early follicles to release oestrogen. Then the little vessels, under vascular endothelial growth factor control, are capped off like tiny oil wells, and blood loss gradually stops.
- And so the new cycle has begun…

5A.1 FOLLICULAR RECRUITMENT AND DEVELOPMENT

Several follicles develop in the early part of the cycle, none of them greater than 10 mm in size, producing (per follicle) tiny amounts of oestrogen. The next big event is the selection of the dominant follicle. Selection of the dominant follicle works in a manner analogous to the way the cuckoo bird lays its eggs. Cuckoos are clever birds. They do not build their own nests, but instead lay their eggs (generally only one per nest) in other bird species' nests. The cuckoo eggs will often be different colours and slightly bigger than the host eggs in the nest. These will be more favoured by the hatching bird and as a consequence will hatch first, so they are fed first over all the other hatchlings in the nest. Then, while the mother or father bird is off scavenging for worms, the newly hatched cuckoo edges all the competitor native eggs out of the nest, so that the cuckoo ends up getting all the nourishment.

The development of a follicle is dependent on the internal formation of the antrum, a fluid-filled space. The fluid is composed of fluid from capillary blood vessels in the theca cells and chemicals secreted by the granulosa cells (gonadotrophins and other trophic hormones, oestrogens and progestogens, antioxidants and growth factors) (40). Murine studies show that the granulosa cells differentiate into two types – those around the internal wall of the follicle (mural cells) and those around the oocyte (the cumulus oophorus). Mural cells produce steroid hormones, while cumulus cells support the oocyte in its growth. Cholesterol is broken enzymatically to produce the sex steroids.

Oocytes are remarkably incompetent; their only ability is to utilise substrate and grow, and grow they do, starting at 20 μm and finishing at 80 μm. Oocytes cannot utilise glucose as an energy source, they cannot make key amino acids and they do not have the enzymes required to break down cholesterol into oestrogen. They rely totally on their associated granulosa cells for pyruvate and adenosine triphosphate (ATP) and for transporting ions, metabolites and amino acids through the gap junctions in the oocyte/granulosa cell membrane (41). Oocytes have some abilities. Cholesterol itself is synthesised in the cumulus cells, and the oocyte is key to this – absence of the oocyte has been shown to suppress cholesterol metabolism, and mural cells produce far less cholesterol than cumulus cells (42). The transforming growth factor β (TGFβ) group, growth differentiation factor 9 (GDF9) and bone morphogenetic protein 15 (BMP15) are all paracrine factors secreted by the oocyte which control follicular development. GDF9 appears to control granulosa cell function and follicle development, while BMP15 is responsible for cumulus cell expansion just before ovulation.

The follicle which becomes the dominant one is generally well advanced in its development compared to all the others. It develops (FSH) receptors on its surface more rapidly than the other follicles; as a result, it is more responsive to circulating FSH (getting more worms, in the cuckoo analogy), and it grows more quickly. As it does so, levels of oestrogen in the blood are increasing, and the feedback loop causes the pituitary to reduce circulating FSH levels. The FSH stimulus to all follicles falls off, but the dominant follicle has many more receptors than the smaller follicles and is able to maintain its size, whereas all the others shrink (being extruded from the nest). Only that one follicle generally continues on to maturity and to ovulation. On ultrasound, this will be seen as the image in Figure 5A.1.

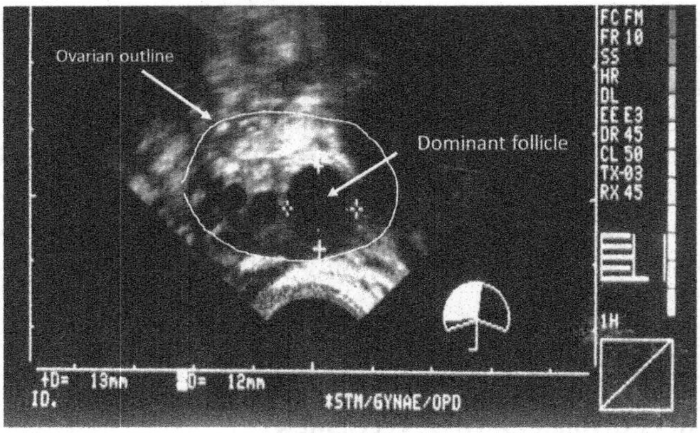

Figure 5A.1 Appearance of a dominant follicle.

Work done by Roger Gosden in Leeds just over 20 years ago demonstrated how the follicle develops responsiveness to FSH. This is illustrated in cartoon form in Figure 5A.2a–c. In essence, as the granulosa cell layer develops, receptors to FSH develop: 100% of mature, multilaminar follicles have receptors to FSH, whereas 40% of 1–2 layer and 0% of primordial follicles have any receptors (43).

The above account is the textbook description of how the dominant follicle develops, describing a linear progression of one follicle to maturation. We know, however, that it often doesn't work like that. In unpublished observations, I performed daily assessment of follicle development in natural cycles over 4 years. Over the 3–4 days of early follicular development, different follicles appeared to be the dominant one, larger than the others, but each day showed a different dominant follicle, until finally one broke ahead of the others, and all the others regressed. This was confirmed for me at an European Society of Human Reproduction and Embryology conference in the early 1990s, when I saw data presented by the Brussels group (never published that I am aware of) showing how follicular dynamics in the ovary are flexible and the follicle which starts off as apparently dominant often recedes, with another taking its place.

Between recruitment of the dominant follicle and ovulation, a number of key endocrinological events take place. Figure 5.1 illustrates in cartoon format how this happens. Gonadotrophin-releasing hormone (Table 5.3) is a decapeptide hormone, synthesised in the arcuate nucleus of the hypothalamus. I like to compare the arcuate nucleus to the first violin leader of the orchestra, driving the beat, the pulses of GnRH. GnRH is released in a pulsatile manner from axonal ends of arcuate nucleus neurones into the pituitary portal system. The origin, the driver of GnRH pulsatility, is not known for certain. It may be self-generating, or it may be stimulated by KNDy cells, referred to in Chapter 1 (Figure 1.1). In several animal models, there appears to be direct axonal contact from the KNDy cells to the arcuate nucleus cells, but this has not been demonstrated in humans (11). GnRH pulsatility is critical to the way that gonadotrophin secretion is regulated. In the early follicular phase, the pulse frequency is every 90 minutes; at midcycle, it is more rapid at every 60–70 minutes and in the luteal phase, it is much slower, every 120–200 minutes or so (1). Gonadotroph cells in the pituitary are of differing subtypes: monohormonal and multihormonal, demonstrated in animal studies (44). The monohormonal subtype responds to GnRH simulation by producing either FSH or LH only; the multihormonal subtype responds to stimulation by producing both FSH and LH. The different type of cell is activated by the different frequencies of the GnRH pulses. A pulse rate of 90 or 120 minutes stimulates the monohormonal cells, while a pulsatility of 60 minutes activates the multihormonal cells.

The above account does not include the roles of the inhibins or activin. It does, however, provide an understandable model to grasp how the endocrine control of ovulation is managed and explains

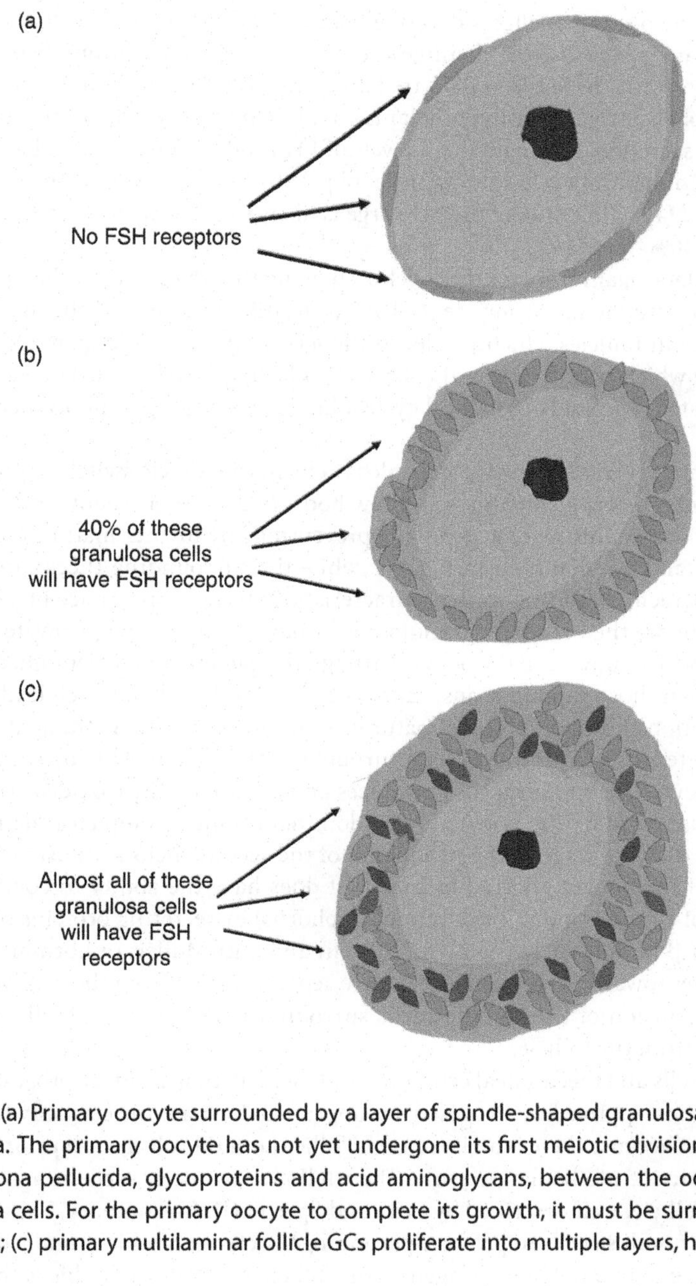

Figure 5A.2 (a) Primary oocyte surrounded by a layer of spindle-shaped granulosa cells demarcated by a basal lamina. The primary oocyte has not yet undergone its first meiotic division; (b) primary unilaminar follicle, zona pellucida, glycoproteins and acid aminoglycans, between the oocyte and a single later of granulosa cells. For the primary oocyte to complete its growth, it must be surrounded by a layer of granulosa cells; (c) primary multilaminar follicle GCs proliferate into multiple layers, hence multilaminar.

how the seemingly unwanted and unneeded FSH midcycle surge seen in Figure 5.5 comes about. Currently, it is not known why this FSH surge occurs, and it does not appear to have any role in the control of the ovulatory process.

5A.2 WORKING TOWARDS OVULATION

The endocrinology that leads to ovulation is fascinating. As the dominant follicle grows, it breaks down cholesterol into more and more oestradiol. This hormone is active at a number of sites throughout the body: it is stimulatory to the endometrium and to the breast; it induces growth and maturation of the oocyte and it has feedback functions within the follicle, as well as to the pituitary and the hypothalamus. It also supports tissue health in skin, bone, hair, the vulva and the vagina.

In addition to the oestrogen effects described above, oestradiol gives feedback to the pituitary and the hypothalamus. Because the dominant follicle has high receptor numbers, it benefits from FSH stimulus even while the levels of FSH are falling, mediated by the rising oestradiol. The rising levels of oestradiol also trigger the hypothalamus, and then the pituitary, to reduce pulsatility frequency of GnRH secretion, leading to high levels of LH production, recognisable as the LH surge. In my experience of natural cycle studies, little 'blips' of LH often occur in the days leading up to the 'true' LH surge (45). The onset of the LH surge is likely to be between 4 a.m. and 8 a.m. (in 48% of cycles, 95% CI 40%–60%) (45).

There are also data, again from natural cycles, showing that the extent of the LH surge is related to the likelihood of producing a competent, that is, a fertilisable oocyte. Follicular fluid (FF) levels of LH were higher in follicles which produced the fertilised oocyte compared with follicles producing an oocyte which wasn't fertilised (14.6 vs 10.4 IU/L, $p = 0.01$) (46,47). Sadly, these studies were never replicated. The study of natural cycle follicles is not of high importance in the minds of researchers.

The LH surge has several effects: the granulosa cells in the follicle switch to producing progesterone (and becomes a corpus luteum – a yellow body, *lutein* means yellow in Latin), and all the other follicles are driven into atresia, generally preventing any further maturation. The granulosa cells also enlarge and develop lipid inclusions, while the surrounding theca-luteal cells become vascular and hyperaemic, leading to the characteristic external appearance of yellowish-redness of luteinisation. Before the LH surge, granulosa cells have been making pyruvate, ATP and cyclic AMP for energy, which move into the oocyte through the gap junctions mentioned earlier (41). LH release breaks down those gap junctions, causing cAMP levels to fall, which leads to suppression of oocyte maturation inhibitor. Oocyte maturation inhibitor (OMI) is thought to be the factor which keeps oocytes in their dormant state throughout the woman's life by suppressing meiotic activity in the oocyte. The LH surge also produces an increase in nitric oxide activity, a promoter of meiotic progression (48). The oocyte in the dominant follicle completes the meiotic process, moving from the dictyate stage to the metaphase of the second meiotic division (49). Despite the 'incompetence' of the oocyte referred to earlier, it does have the ability to control the development of a group of cells around it (the cumulus oophorus) by secreting proteins of the transforming growth factor β family, such as activin, inhibin and anti-Mullerian hormone. As the oocyte matures and moves towards ovulation, the oocyte separates itself from the wall of the follicle and also promotes expansion of the cumulus cells surrounding the oocyte. Finally, the oocyte ovulates, being freed from the follicle.

The granulosa cells and theca-luteal cells, under LH stimulation, produce progesterone in greater amounts as ovulation approaches. Progesterone appears to increase the elasticity of follicular cells, which is important because there is an increase in follicular fluid volume just prior to ovulation, yet no major increase in intrafollicular pressure. There follows an orderly succession of steps to free the oocyte from the follicle, well described by Fritz and Speroff (1). In summary, FSH, LH and progesterone stimulate granulosa cells to produce plasminogen activators which produce plasmin, which triggers collagenase to break down collagen in the follicle wall. (Possibly this is the reason for the FSH surge noted in Figure 5.5.) The LH surge also stimulates the production of several eicosanoids, mostly prostaglandin E_2, which in turn bring about release of proteolytic enzymes acting on the follicle wall and possibly also mediating angiogenesis and hyperaemia (as an inflammatory type response). Smooth muscle cells have also been identified in the stroma of the ovary, and the prostaglandins may stimulate these to contract to further augment ovulation and extrusion out of the follicle, hopefully into the waiting fimbriae of the fallopian tube. This is an important therapeutic point – inhibitors of prostaglandins are reputed to prevent ovulation and thereby cause infertility (50). Fuller discussion of this occurs elsewhere (Section 7.5.6, 'Prescribed and Over-the-Counter Medications', in Chapter 7).

In this description of ovulation, I have deliberately avoided describing the other factors that are involved in the process – activin, the inhibins – sticking to the core functions for simplicity.

Those additional factors are important, but they cannot be manipulated therapeutically to the same extent. Note also the words 'negative feedback' have not been included anywhere to this point – to avoid the confusion that such concepts wrongly introduce.

5A.3 SUMMARY

This appendix presents, a model for understanding how ovulation is controlled, how a single follicle gets recruited and selected and how, simultaneously with ovulation, a number of key oocyte functions occur which are critical to fertilisation.

REFERENCES

1. Fritz MA, Speroff L. *Clinical Gynecologic Endocrinology and Infertility.* Lippincott Williams & Wilkins; Philadelphia, Pennsylvania 2010.
2. Oliver A, Awala AO, Skidmore C, Cahill DJ. Recognition and reduction of the impact of inheritable causes of ovarian failure. *Journal of Obstetrics and Gynaecology: Journal of the Institute of Obstetrics and Gynaecology.* 2009;29(5):445.
3. Rotterdam ESHRE/ASRM-Sponsored PCOS Consensus Workshop Group Revised 2003 consensus on diagnostic criteria and long-term health risks related to polycystic ovary syndrome. *Fertility and Sterility.* 2004;81(1):19–25.
4. Dumesic DA, Oberfield SE, Stener-Victorin E, Marshall JC, Laven JS, Legro RS. Scientific statement on the diagnostic criteria, epidemiology, pathophysiology, and molecular genetics of polycystic ovary syndrome. *Endocrine Reviews.* 2015;36(5):487–525.
5. Homburg R. Involvement of growth factors in the pathophysiology of polycystic ovary syndrome. *Gynecological Endocrinology: Official Journal of the International Society of Gynecological Endocrinology.* 1998;12(6):391–7.
6. Nestler JE, Jakubowicz DJ. Decreases in ovarian cytochrome P450c17 alpha activity and serum free testosterone after reduction of insulin secretion in polycystic ovary syndrome. *New England Journal of Medicine.* 1996;335(9):617–23.
7. Pasquali R. Metabolic syndrome in polycystic ovary syndrome. *Frontiers of Hormone Research.* 2018;49:114–30.
8. Norman RJ, Dewailly D, Legro RS, Hickey TE. Polycystic ovary syndrome. *Lancet (London, England).* 2007;370(9588):685–97.
9. Torre DL, Falorni A. Pharmacological causes of hyperprolactinemia. *Therapeutics and Clinical Risk Management.* 2007;3(5):929–51.
10. Fernandez A, Karavitaki N, Wass JA. Prevalence of pituitary adenomas: A community-based, cross-sectional study in Banbury (Oxfordshire, UK). *Clinical Endocrinology.* 2010;72(3):377–82.
11. Lehman MN, Coolen LM, Goodman RL. Minireview: Kisspeptin/neurokinin B/dynorphin (KNDy) cells of the arcuate nucleus: A central node in the control of gonadotropin-releasing hormone secretion. *Endocrinology.* 2010;151(8):3479–89.
12. Cahill DJ, Wardle PG. Management of infertility. *BMJ: British Medical Journal.* 2002;325(7354):28–32.
13. National Collaborating Centre for Women's and Children's Health. *National Institute for Health and Clinical Excellence: Guidance. Fertility: Assessment and Treatment for People with Fertility Problems.* London: Royal College of Obstetricians & Gynaecologists National Collaborating Centre for Women's and Children's Health. 2013.
14. Rossing MA, Daling JR, Weiss NS, Moore DE, Self SG. Ovarian tumors in a cohort of infertile women. *New England Journal of Medicine.* 1994;331(12):771–6.
15. Brinton LA, Lamb EJ, Moghissi KS, Scoccia B, Althuis MD, Mabie JE et al. Ovarian cancer risk after the use of ovulation-stimulating drugs. *Obstetrics and Gynecology.* 2004;103(6):1194–203.

16. National Collaborating Centre for Women's and Children's Health. *National Institute for Health and Clinical Excellence: Guidance. Fertility: Assessment and Treatment for People with Fertility Problems*. London: Royal College of Obstetricians & Gynaecologists National Collaborating Centre for Women's and Children's Health; 2013.
17. Cahill DJ, O'Brien K. Polycystic ovary syndrome (PCOS): Metformin. *BMJ Clinical Evidence*. 2015;2015.
18. Schwetz V, Eisner F, Schilcher G, Eller K, Plank J, Lind A et al. Combined metformin-associated lactic acidosis and euglycemic ketoacidosis. *Wiener Klinische Wochenschrift*. 2017;129(17–18):646–9.
19. Palomba S, Falbo A, La Sala GB. Metformin and gonadotropins for ovulation induction in patients with polycystic ovary syndrome: A systematic review with meta-analysis of randomized controlled trials. *Reproductive Biology and Endocrinology: RB&E*. 2014;12:3.
20. Tso LO, Costello MF, Albuquerque LE, Andriolo RB, Marjoribanks J, Macedo CR. Metformin treatment before and during *in vitro* fertilization or intracytoplasmic sperm injection in women with polycystic ovary syndrome: Summary of a Cochrane review. *Fertility and Sterility*. 2015;104(3):542–4.
21. Given JE, Loane M, Garne E, Addor MC, Bakker M, Bertaut-Nativel B et al. Metformin exposure in first trimester of pregnancy and risk of all or specific congenital anomalies: Exploratory case-control study. *BMJ (Clinical Research Ed)*. 2018;361:k2477.
22. Jenkins JM, Davies DW, Anthony F, Wood P, Gadd SG, Watson RH et al. The detrimental influence of functional ovarian cysts during *in-vitro* fertilization cycles. *Human Reproduction (Oxford, England)*. 1992;7(6):776–80.
23. Porchet HC, le Cotonnec JY, Loumaye E. Clinical pharmacology of recombinant human follicle-stimulating hormone. III. Pharmacokinetic-pharmacodynamic modeling after repeated subcutaneous administration. *Fertility and Sterility*. 1994;61(4):687–95.
24. Cahill DJ, Wardle PG, Harlow CR, Hunt LP, Hull MG. Expected contribution to serum oestradiol from individual ovarian follicles in unstimulated cycles. *Human Reproduction (Oxford, England)*. 2000;15(9):1909–12.
25. Yumusak OH, Kahyaoglu S, Pekcan MK, Isci E, Cinar M, Tasci Y. Does intrauterine insemination timing matter for achieving pregnancy during ovulation induction using gonadotropins? A retrospective cohort study. *Journal of the Chinese Medical Association: JCMA*. 2017;80(6):366–70.
26. Shalev E, Leung PC. Gonadotropin-releasing hormone and reproductive medicine. *Journal of Obstetrics and Gynaecology Canada: JOGC*. 2003;25(2):98–113.
27. Ron-El R, Raziel A, Schachter M, Strassburger D, Kasterstein E, Friedler S. Induction of ovulation after GnRH antagonists. *Human Reproduction Update*. 2000;6(4):318–21.
28. Fleming R, Jamieson ME, Hamilton MP, Black WP, Macnaughton MC, Coutts JR. The use of GnRH analogues in combination with exogenous gonadotropins in infertile women. *Acta Endocrinologica Supplementum*. 1988;288:77–84.
29. Out HJ, Rutherford A, Fleming R, Tay CC, Trew G, Ledger W et al. A randomized, double-blind, multicentre clinical trial comparing starting doses of 150 and 200 IU of recombinant FSH in women treated with the GnRH antagonist ganirelix for assisted reproduction. *Human Reproduction (Oxford, England)*. 2004;19(1):90–5.
30. Zollner U, Neumann C, Zollner KP. Probability of success of conservative treatment of infertility: Exact timing as an important prognostic factor. *Journal of Reproductive Medicine*. 2013;58(5–6):205–11.
31. Bhattacharya S, Harrild K, Mollison J, Wordsworth S, Tay C, Harrold A et al. Clomifene citrate or unstimulated intrauterine insemination compared with expectant management for unexplained infertility: Pragmatic randomised controlled trial. *BMJ (Clinical Research Ed)*. 2008;337:a716.

32. HFEA. HFEA website: HFEA; 2018. Available from: https://www.hfea.gov.uk/treatments/explore-all-treatments/intrauterine-insemination-iui/.
33. Elnaggar EA, Elwan YA, Ibrahim SA, Abdalla MM. Hormonal changes after laparoscopic ovarian diathermy in patients with polycystic ovarian syndrome. *Journal of Obstetrics and Gynaecology of India*. 2016;66(Suppl 1):528–33.
34. Luz R, Barros J, Aguiar A, Rodrigues C, Soares AP, Nunes J et al. Short and long-term efficacy of laparoscopic ovarian diathermy in women with polycystic ovary syndrome. *Acta Medica Portuguesa*. 2016;29(7–8):441–8.
35. Farquhar C, Brown J, Marjoribanks J. Laparoscopic drilling by diathermy or laser for ovulation induction in anovulatory polycystic ovary syndrome. *Cochrane Database of Systematic Reviews*. 2012(6):Cd001122.
36. Fahy UM, Cahill DJ, Wardle PG, Hull MG. Osteoporotic fractures in young amenorrheic women. *Acta Obstetricia et Gynecologica Scandinavica*. 1994;73(5):417–9.
37. Azziz R. Polycystic ovary syndrome. *Obstetrics and Gynecology*. 2018;132(2):321–36.
38. Unfer V, Carlomagno G, Papaleo E, Vailati S, Candiani M, Baillargeon JP. Hyperinsulinemia alters myoinositol to d-chiroinositol ratio in the follicular fluid of patients with PCOS. *Reproductive Sciences (Thousand Oaks, Calif)*. 2014;21(7):854–8.
39. Kurzthaler D, Hadziomerovic-Pekic D, Wildt L, Seeber BE. Metformin induces a prompt decrease in LH-stimulated testosterone response in women with PCOS independent of its insulin-sensitizing effects. *Reproductive Biology and Endocrinology: RB&E*. 2014;12:98.
40. Basuino L, Silveira CF, Jr. Human follicular fluid and effects on reproduction. *JBRA Assisted Reproduction*. 2016;20(1):38–40.
41. Su YQ, Sugiura K, Eppig JJ. Mouse oocyte control of granulosa cell development and function: Paracrine regulation of cumulus cell metabolism. *Seminars in Reproductive Medicine*. 2009;27(1):32–42.
42. Su YQ, Sugiura K, Wigglesworth K, O'Brien MJ, Affourtit JP, Pangas SA et al. Oocyte regulation of metabolic cooperativity between mouse cumulus cells and oocytes: BMP15 and GDF9 control cholesterol biosynthesis in cumulus cells. *Development (Cambridge, England)*. 2008;135(1):111–21.
43. Oktay K, Briggs D, Gosden RG. Ontogeny of follicle-stimulating hormone receptor gene expression in isolated human ovarian follicles. *Journal of Clinical Endocrinology and Metabolism*. 1997;82(11):3748–51.
44. Lloyd JM, Childs GV. Differential storage and release of luteinizing hormone and follicle-releasing hormone from individual gonadotropes separated by centrifugal elutriation. *Endocrinology*. 1988;122(4):1282–90.
45. Cahill DJ, Wardle PG, Harlow CR, Hull MG. Onset of the preovulatory luteinizing hormone surge: Diurnal timing and critical follicular prerequisites. *Fertility and Sterility*. 1998;70(1):56–9.
46. Cahill D, Harlow C. Does a reduction in the mid-cycle LH surge quality affect fertilisation of oocytes *in vitro*? *British Journal of Obstetrics and Gynaecology*. 1997;104(6):638–9.
47. Verpoest WM, Cahill DJ, Harlow CR, Hull MG. Relationship between midcycle luteinizing hormone surge quality and oocyte fertilization. *Fertility and Sterility*. 2000;73(1):75–7.
48. Voznesens'ka T, Blashkiv TV. Effect of NO on meiotic maturation of the oocytes in mice *in vitro*. *Fiziolohichnyi zhurnal (Kiev, Ukraine: 1994)*. 2003;49(1):100–3.
49. Tsafriri A, Pomerantz SH. Oocyte maturation inhibitor. *Clinics in Endocrinology and Metabolism*. 1986;15(1):157–70.
50. Duffy DM. Novel contraceptive targets to inhibit ovulation: The prostaglandin E_2 pathway. *Human Reproduction Update*. 2015;21(5):652–70.

Damage to the uterus, the fallopian tubes and the ovaries

6

OVERVIEW

This chapter covers those conditions which cause distortion and damage to the uterus, the fallopian tubes and the ovaries, and includes options for management.

6.1 THE UTERUS

The pathological conditions affecting the uterus can be divided into: congenital structural abnormalities and acquired conditions, the latter divided into myometrial pathology (largely fibroids and adenomyosis) and endometrial pathology (caused by surgery, infection and polyps).

6.1.1 Congenital uterine abnormalities

Congenital uterine abnormalities develop from failure of normal fusion of the embryological Mullerian tract. These congenital abnormalities vary from almost imperceptible indentations of the uterine fundus to the finding of a double uterus, cervix and vagina. Minor mid-line downward indentation is such a common finding that it can be considered a variation of normal. Between normal and the extreme finding of a double uterus, cervix and vagina, there is a wide range of abnormalities.

The incidence of these uterine anomalies is worth a moment's consideration – published meta-analyses suggest an incidence in infertile women of 3.4%–8.0%, and in the recurrent miscarriage group up to 12.6%–18.2% (1). Having worked in a tertiary gynaecology centre for almost 30 years, I am very surprised by these figures, as I rarely observed congenital uterine abnormalities.

The crucial issue for the patient is: What influence, if any, do these abnormalities have on fertility and the ability to carry a pregnancy?

These abnormalities will have an impact on fertility and pregnancies if the implanting embryo has little or no stroma or myometrium to support its implantation. This is what happens with complete bicornuate, and completely septate, uteri. The layer of endometrium on the septum or the distal end of the bicornuate uterus has very little stromal tissue underneath it. Decidualisation, 'the response of the endometrium to the invading trophoblast', is also dependent on stromal tissue quality (2). Normally, the trophoblast releases growth factors (tumour necrosis factor α [TNFα] and cytokines, amongst others) to stimulate the maternal blood vessels in the stromal tissue to grow towards the trophoblast, whereupon the trophoblast invades those blood vessels (4). Colour Doppler blood flow studies in the uterine abnormalities under discussion have shown diminished or absent blood flow in large sections of septa and 'poor decidualisation and placentation due to the reduced amounts of (stromal) connective tissue' (3). This means there is no onward facility for blood vessel and glandular development.

It is therefore surprising that material differences in outcomes between women with congenital anomalies and controls have not been found in meta-analyses of natural or assisted reproduction cycles when the cycles were analysed separately. When all cycles were combined and compared with normal controls, there was a trend to significance for structural abnormalities causing poor pregnancy outcomes, but the evidence was underwhelming (1). Nevertheless, anatomical abnormalities are a lure to surgical correction, and the patient should know what the likely benefits will be. In their meta-analysis of hysteroscopic resection of anatomical abnormalities, Venetis and Papadopoulos found no advantage in achieving pregnancy with surgical correction, although the miscarriage rate *was* lower in the treated group (1).

My view is that most uterine anomalies contribute little to fertility problems, and surgery offers little to improve the likelihood of getting pregnant. However, removal of septa if there is a history of recurrent miscarriages may well be helpful, provided other causes for miscarriage have been excluded.

6.2 UTERINE FIBROIDS

Fibroids, termed myomas or leiomyomas, but more correctly leiomyomata, are generally benign uterine tumours. In 28 years of regularly performing myomectomies, I encountered only one woman with cancerous fibroids. Fibroids grow slowly *in vivo* and are found within the extracellular matrix. They arise from the myometrium. Fibroids contain smooth muscle cells and extracellular fibrous tissue comprising collagen, proteoglycan and fibronectin. The smooth muscle cells of the fibroid are monoclonal (arising from single clone of a smooth muscle cell); up to 50% of fibroids will be chromosomally abnormal, apparently varying by the site of the fibroid (5). Fibroids have a dense connective tissue capsule, visible on ultrasound, on MRI and at surgery. Correct and careful surgical opening will allow access to a 'relatively' blood-free plane of cleavage between the capsule and the fibroid, which enables removal of the fibroid, with minor bridging arterioles being treated for haemostasis. Surprisingly, in practice, there is rarely a feeding artery at the base of the fibroid. Fibroids can be intramural, found completely within the myometrium, or they may be submucosal, partially or entirely protruding into the uterine cavity and sometimes pedunculated. They may also be partially or entirely outside of the myometrium, subserosal.

Fibroids are common and estimates of their frequency vary from 30% on clinical assessment and 50% on ultrasound (6). When hysterectomy samples were examined (hysterectomies for all reasons: heavy periods, pain, etc.), fibroids were found in more than 70% of samples (7). Fibroids are generally found in women of reproductive years, giving rise to the theory that they are initiated and/or maintained by stimuli from oestradiol and progesterone (8). It is more complex than that, of course, and various proteins and growth factors are likely to be implicated. Other aetiological mechanisms have been postulated, including infection-related oncogenesis, abnormal cervical cytology, the use of genital talc, the use of intrauterine devices and sexually transmitted diseases. The National Institute of Environmental Health Sciences Uterine Fibroid Study surveyed more than 1000 women, examining these putative factors. The study found no link with most infections, including latent Chlamydia, but it found an inverse association between fibroids and abnormal cervical cytology (9). This was a well-conducted study which helped to clarify important questions, and the inverse relationship of abnormal cervical cytology and fibroids prompts further speculation. Human papilloma virus (HPV), the most common cause of abnormal cervical cytology, is known to be inserted at chromosome 12q14-15 in cervical cancer. That region on chromosome 12 is also the region of a translocation seen in fibroid tumours which have reached a larger size. The HPV insertion site is 50–100,00 base pairs (that's very close) from the HMGA-2 gene which can be overexpressed in fibroids. The inverse relationship described above suggests that HPV insertion may stabilise the region and reduce translocations, thus limiting fibroid growth. This area of work needs further exploration to test the hypothesis, but, if fruitful, there is the exciting potential that it could generate a vaccine to protect against fibroids in at-risk women.

6.2.1 Effect of fibroids on fertility

There is a prevailing belief that fibroids adversely affect fertility. One systematic review found data suggesting that submucosal fibroids were associated with lower than normal ongoing pregnancy rates (14% vs 30%, OR 0.44; 0.28–0.70), but that women with intramural fibroids had a nonsignificant reduction in pregnancy rates, 37% vs 41% (OR 0.84; 0.74–0.95) (10). Pritts and colleagues found similar results (11). The evidence of the effect of fibroids which do not distort the uterine cavity is less strong, although a meta-analysis examining over 6000 IVF treatment cycles showed that the clinical pregnancy and live birth rates were reduced by 15% and 21%, respectively, in women with non-cavity distorting fibroids (RR 0.85, 95% CI: 0.77–0.94 and RR 0.79, 95% CI: 0.70–0.88) (12).

If it is the case that fibroids which impinge into the uterine cavity (submucosal and intramural fibroids) cause a reduction in fertility, is there any evidence for improvement after treatment?

There was some support for intervention from a retrospective study of 31 infertile women having myomectomy – 11 carried a pregnancy following surgery (13). However, two systematic reviews were less encouraging, finding mostly insignificant increases in pregnancy and live birth rates in those undergoing surgery, particularly in intramural fibroids; removal of submucous myomas does appear to improve fertility (11,14). A 2012 Cochrane review concluded that there was not enough evidence to justify myomectomy to improve fertility (15). It's worth noting that even this Cochrane review had only three studies to work with, and an improvement in fertility outcomes was reported as an outcome in only one of those. Clearly, more work is needed in this area to clarify an ongoing debate.

So what advice should be given to women who present with infertility and who have one or more fibroids? It seems sensible to limit myomectomy to the submucosal fibroid variety. For intramural fibroids, I would only consider surgery if all other causes have been excluded. This recognises the impact that surgery has on people's lives and that the surgery is not without risk, something which will be discussed in the surgery section (Section 6.2.4, 'Surgical Treatment').

6.2.2 Medical options

Medical options include the use of gonadotrophin-releasing hormone agonist drugs (GnRHa drugs), or ulipristal, a selective progesterone receptor modulator (SPRM). These not only have a place in the medical management of fibroids, but they are also frequently used prior to myomectomy or hysterectomy to shrink fibroids, making the surgical procedure safer and easier. The GnRHa family (e.g. triptorelin, goserelin, leuprolide acetate) all shrink fibroids effectively (16). The maximum effect of fibroid shrinkage is likely to be achieved within 3–4 months. Treatment can continue beyond that to maintain the smaller size if the woman needs to lose weight or improve respiratory function or haemoglobin before surgery. The negative effect of long-term GnRHas on bone density means that treatment should not go beyond 6 months in the absence of 'add-back' therapy. Add-back therapy is a medication (generally some form of hormone replacement therapy [HRT]), to offset and minimise the menopausal side effects that women experience on GnRHa treatment, particularly bone loss. Tibolone is converted into three major metabolites, two of which have oestrogenic effects, and one which has progestogenic and androgenic effects (17). It is the most effective drug for managing bone loss and vasomotor symptoms associated with GnRNa use (18).

Several papers suggest that ulipristal may have a wider use beyond just surgical pre-operative shrinkage. In two studies containing 209 and 134 women, serial 3-monthly courses of ulipristal were effective in shrinking fibroids and reducing heavy periods in approximately 75% of women over a longer period of time (19,20). In 2018, notable problems with ulipristal administration came to light, principally severe liver complications. The US Food and Drug Administration has refused to licence its use for treating fibroids. The European Medicines Agency (EMA) (in its recommendations of August 2018) restricts its use to one treatment course per patient, unless the patient is not eligible for surgery and would benefit from it. This will certainly curtail its use, and the EMA's additional conditions will hopefully prevent any further adverse problems (21). These conditions are:

- It should not be used if there is pre-existing liver disease.
- Liver function tests should be done before, during and after treatment.
- The patient should be given written information about the need for liver monitoring and instructions to contact their doctor if they develop signs of liver problems.

6.2.3 Nonoperative treatment: Uterine artery embolisation

Uterine artery embolisation of fibroids offers an attractive, nonsurgical treatment for fibroids and is particularly helpful when a woman is reluctant or unfit to have surgery. The procedure is generally done as a day case procedure, with minimal impact on lifestyle and rapid return to normal daily activities, including work. The procedure was developed to treat the heavy uterine bleeding and anaemia associated with fibroids, and a 90% success rate has been demonstrated (22).

The fibroid embolisation technique uses direct injection into the blood stream of small particles which block the blood supply of the fibroid. Access is usually achieved through the femoral artery, and then, under fluoroscopic control, the arterial supply to the fibroid is identified. It is often the distal horizontal segment of the uterine artery, beyond the cervico-vaginal branch (23). A micro-injection of embolising particles (such as nonspherical polyvinyl alcohol particles or calibrated Tris acryl microspheres >500 μm diameter) are injected. If blood flow stasis is not achieved, further doses are administered. While the effect of embolisation is to shrink the fibroid(s), shrinkage will generally not exceed 50% of the former size, and a moderate proportion (15%–30%) will require surgery, either myomectomy or hysterectomy (24).

Reports of minor and major complications have emerged over time. These are postprocedure pain (severe enough to require readmission and strong analgesics), infection, premature ovarian failure (probably as a result of imprecise placement of the uterine catheters and secondary embolisation of the ovary as well as the uterus) and secondary amenorrhoea due to endometrial atrophy or intrauterine adhesions. Most reports are too small in numbers to provide useful data on complications, but the Health Technology Assessment (HTA) has data on over 1100 women (650 of whom had endoscopic embolisation compared with 460 who had hysterectomy for fibroids) (25). Severe or major complications occurred in 52 (11%) of women who had a hysterectomy as their primary treatment, and 25 (4%) of women who had embolisation – higher in the hysterectomy group, but not significantly so. In the embolisation group, permanent amenorrhoea in the under-40s occurred once (0.2%); in the over-40s it occurred in nine women (1%). Septicaemia or emergency myomectomy or hysterectomy occurred in 17 women (2%). Frustratingly, the report does not differentiate between these three very serious conditions, making it difficult to give good information to prospective patients. Postprocedure pain and infection were lower in women having embolisation – under 10%, compared with 19% in women having hysterectomy ($p < 0.01$). The same study looked at comparative complications rates for hysterectomy and embolisation in the management of fibroids. It found the major complication rate was 14% and 4%, respectively (blood transfusion, septicaemia or structural damage caused), and severe complications occurred in 1% and 0.2%, respectively (severe meaning pulmonary embolus and organ failure) (25). To prevent and manage these possible complications, the practice in our hospital is to admit women for 24 hours to manage pain symptoms, generally using rectal diclofenac, but occasionally patient-controlled anaesthesia. Pain results from the necrosis of the fibroid(s) and can be severe. Prophylactic antibiotics appropriate to gram-negative bacteria are given. Infection may occur at the injection site, but can also occur in the uterus, leading to hysterectomy – in up to 5% of cases in one small series (22), but see the HTA report earlier of 2% (25). It is important to counsel women that the most they can expect in fibroid size reduction is 33%–50%.

6.2.3.1 Effect of uterine artery embolisation on fertility

Whether embolisation of fibroids improves the chances of achieving a pregnancy has been widely debated. The more conservative view is that the uterine muscle wall will be weakened by the spread of embolisation material beyond the fibroid; however, several successful pregnancies have been reported following embolisation. There are no convincing large studies to persuade us either way – a 2017 paper was supportive of pregnancy following embolisation, and there were no complications, but only 15 women were included (26). Pregnancy after fibroid embolisation is not generally recommended because of the dangers of the embolisation effect on the myometrium in blood supply and integrity leading to implantation and placental blood disorders and, less likely, of uterine rupture (27). In addition, some reports suggest that such pregnancies are complicated by fetal growth retardation or by miscarriage (22,26).

6.2.4 Surgical treatment

In undertaking any surgical procedure for fibroids, by whatever route, it is salutary to remember that myomectomy was compared with controls (no operative procedure) as part of a Cochrane review, and no clear conclusion was made as to either the impact on clinical pregnancy rates or miscarriage (15).

There are two surgical approaches for the removal of fibroids – an abdominal and a vaginal approach. The vaginal hysteroscopic method can be delivered in the outpatient/office setting, and skilled operators can remove a small submucous fibroid with ease in that environment. Given that submucous fibroids have the best-quality evidence for a negative effect on fertility, hysteroscopic myomectomy clearly has a role. One longitudinal study of 960 women who had hysteroscopic myomectomy examined their outcomes: the biggest fibroid removed was 7 cm, and there were seven complications (in 960 women) (<1%) (pyrexia [three], postop bleeding [one], uterine perforation [one], and pulmonary oedema [two]) (28) Those are very impressive data, and hysteroscopic removal of a 7-cm fibroid would be quite a challenge.

When planning for hysteroscopic myomectomy, the use of a GnRH agonist is advisable if time permits. Limits to the size of fibroids removed are stricter than those in laparoscopic or open surgery – ideally they should be less than 4 cm in diameter with at least 50% of the fibroid in the cavity of the uterus (29). In preparing for the procedure, the cervix must be dilated to permit a hysteroscope of 8–10 mm in diameter to accommodate an operating channel. Distension fluid will be required. In the past, glycine, a hypotonic solution which allowed the use of monopolar diathermy, was the fluid of choice, but it occasionally caused aqueous intoxication with pulmonary oedema. Newer resectoscope equipment uses bipolar diathermy and normal saline, which, if it gets absorbed, is less likely to cause fluid overload.

6.2.4.1 Laparoscopic (abdominal) myomectomy

Minimally invasive surgery is increasing in popularity and in demand as patients become increasingly aware of its advantages – a shorter period in hospital, and generally a shortened postoperative recovery time. Within the United Kingdom, it is likely that it is being used increasingly, but accurate data about its use is not available. A UK survey of all consultant-level gynaecologists undertaken in 2015 had a low (22%) response rate. The authors considered that the sample reflected the consultant body in age, years as a consultant and gender. Of the 243 who performed myomectomies, 74% were still performing open laparotomy for fibroid removal, while 32% did laparoscopic removal (30). Over the past 20 years, in the United States, there has been a definite shift in the proportion of myomectomies being undertaken by the open versus the laparoscopic route. In 1995, 95% of myomectomies were undertaken using the open route, but in 2014, only 10% were undertaken by that route (31). During that time, laparoscopic myomectomy rose to about one-third of all myomectomy cases. It is interesting that both countries have reached that level of usage (of 30%) for the laparoscopic approach for myomectomy. Concerns about sarcoma spread, and even more so about morcellation injuries (injury to the bowel and other structures from the device used to shatter the fibroid), have led to the adoption of the mini-laparotomy technique (a small incision in the abdomen to facilitate fibroid removal) in combination with laparoscopic removal of the fibroid, as described very well by Glasser (32).

Laparoscopic myomectomy should only be attempted if the fibroid numbers are four or fewer to minimise multiple uterine incisions, and the largest fibroid should not exceed 12 cm on preoperative imaging (33). Laparoscopic removal has its own challenges with entry injuries but also with the need to remove the resected fibroids using a morcellator. Bowel injuries from morcellator use can be minimised by undertaking the technique within the extraction bag. There is, however, some concern about the potential for spillage of any sarcomatous tissue within a fibroid. The available data gives a wide range for sarcomatous change in fibroids, 1 in 500–2000 fibroids (34). In their series of 514, Bean and colleagues found no fibroid had evidence of sarcomatous change (35).

One interesting suggested combination approach is to embolise very large fibroids and then remove the 'shrunken' fibroid laparoscopically. Expected success rates (safe uncomplicated removal laparoscopically) are high, with 512 of 514 planned procedures being carried out (35). In this study, complication rates were low – 18 women had severe complications (blood loss >1000 mL [n = 15], bowel injury [n = 1], bladder injury [n = 1], and small bowel obstruction secondary to port site hernia [n = 1]) (35).

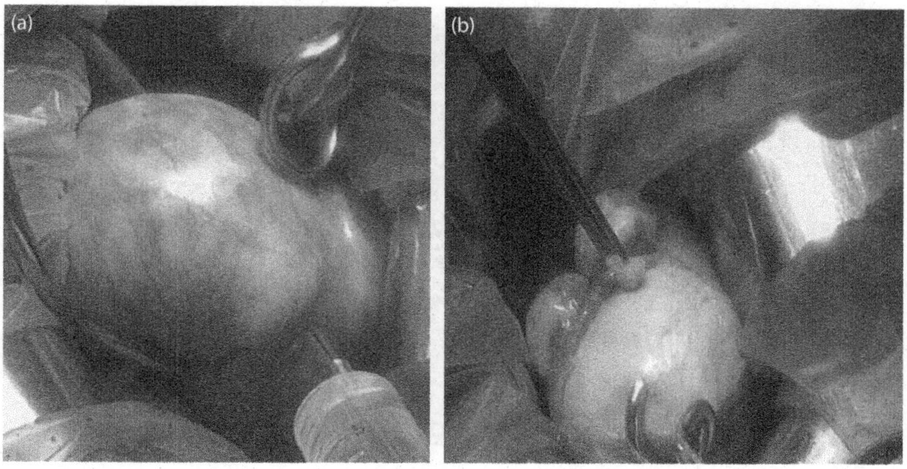

Figure 6.1 Nontouch technique with (a) injection of vasopressin and (b) gentle dissection using moistened pledgets to aid dissection of the fibroid out of its capsule.

As with all fibroid surgical interventions, the use of GnRH agonists prior to surgery and the use of vasopressin intra-operatively reduces blood loss.

At the other end of the spectrum is open laparotomy. From the three studies included in the Cochrane review of myomectomy, there was no clear evidence as to which surgical approach (open or laparoscopy) gave a better fertility outcome, defined as an increase in clinical pregnancy or live birth rates or lowered miscarriage rates (15).

In my hospital, although laparoscopic fibroid removal is used, we have continued to use abdominal myomectomy for fibroids, largely because the presenting cohort of women tend to have complex presentations, with multiple fibroids. Our practice was described and reviewed in 2016 (36). Out of 45 cases, most women had little blood loss (average blood loss 467 mL; average fall in haemoglobin, 20 g/L. Two (4%) women lost more than 1000 mL of blood). This is no different from the 2% blood less reported in laparoscopic surgery (35). Average hospital stay was 3 days. The use of preoperative ulipristal or GnRH agonist and liberal intraoperative use of vasopressin for haemostasis are similar to the laparoscopic approach. Concerted efforts are made to minimise the formation of adhesions, using a nontouch technique (Figure 6.1) and Ringer's lactate for tissue irrigation (37). No mechanical or similar ligatures are used. Open laparotomy can be a prolonged procedure for patient and staff when removing multiple fibroids (Figure 6.2).

The quoted data for risk of intraoperative hysterectomy for uncontrollable bleeding during fibroid surgery vary from 1 in 200 cases, or 0.5% to just over 4%. Four percent was found amongst a cohort of 3300 women by Subramanian and Clark (38), but in all my years of operating, often on complex cases, although I have sometimes come close to it, I have never had to perform an emergency hysterectomy. Nevertheless, women should be carefully counselled about the risk. A hysterectomy in these circumstances would be life changing, depending on the age and the life plans of the woman. When fibroid removal is to assist fertility and bearing a child, hysterectomy is obviously devastating. The patient should be given data on the surgeon and clinic's surgical outcomes over several years to help in making an informed decision.

6.2.5 Adenomyosis

On ultrasound and on MRI scanning, adenomyosis is often mistaken for fibroids, which is why it is discussed here. Adenomyosis is a fascinating condition, wherein endometrial tissue is found within the myometrium. It is like endometriosis (see Section 7.1.1 in Chapter 7), where endometrial tissue is found *outside* the uterus. Adenomyosis has two principles theories of pathogenesis: the first is

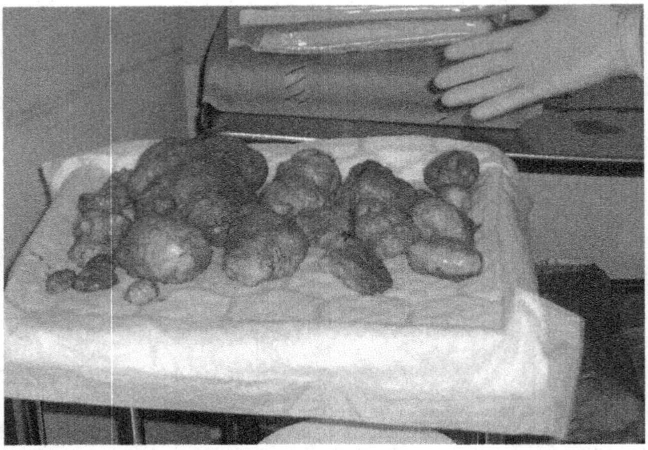

Figure 6.2 Multiple myomectomy of almost 30 fibroids, taking 7 hours of surgical time.

invasion of the endometrium through downgrowth and invagination of the basal endometrium layer into the myometrium through junctional zones (normally a barrier) which are altered or absent; the second is that metaplasia occurs in cell rests in ectopic intramyometrial endometrial tissue (39). On surgical exploration and examination, the characteristic fibroid capsule of a fibroid is not evident. Removal of an adenomyotic mass is difficult because of the absence of a distinct plane of cleavage and extensive invasion of the abnormal tissue into the substance of the myometrium. My practice when encountering this condition in surgery is to excise whatever is feasible and close the myometrium to achieve haemostasis – and terminate the procedure as smoothly as possible.

The effect of adenomyosis on fertility is speculative and there is no clear direction from the available evidence. One radiological study found that 14 of 26 infertile women (primary and secondary) had evidence of adenomyosis (40), but association is not causative evidence. It is often the case that the finding of any condition in women who are infertile is sufficient to generate a desire to treat. However, to quote Devlieger et al., 'uterine adenomyosis remains a fairly frequent and debilitating disease that will be encountered with increasing incidence in the (aging) infertile female population', but there is no clear evidence that adenomyosis is causative in infertility or that treatment improves fertility (41). One systematic review of surgery, with and without prolonged GnRHa therapy for adenomyosis, appears encouraging at first sight. Spontaneous pregnancy rates of 41% in the surgery with GnRHa group were achieved, compared with 15% the surgery-alone group (42). However, the overall figures in the surgery with GnRHa groups were small – 11 out of a total of 27 became pregnant, while in the surgery-only group, n = 187, 28 had a spontaneous pregnancy, that is, 15%. Once again, here is an area which should be investigated more enthusiastically.

6.3 INTRAUTERINE ADHESIONS AND POLYPS

Adhesions (or synechiae) within the uterine cavity (usually called Asherman's syndrome) are rarely congenital and tend to occur after infection or after over vigorous curettage. The infections which predispose to its occurrence include tuberculosis and possibly schistosomiasis (43), and these are reportedly more likely to be found in women in the Indian subcontinent and sub-Saharan Africa (though schistosomiasis is widely endemic in other developing countries, too). In developed countries, clinicians need to be aware of these conditions when dealing with women who have moved from these areas. The conditions predisposing to the development of intrauterine adhesions post-surgery are when the uterus is soft and when the uterine lining is less firmly attached to the stroma. This occurs usually in pregnancy or just after pregnancy, so overvigorous curettage after a miscarriage or a retained placenta can lead to adhesion formation. The use of uterine compression sutures (such as the B-Lynch technique) has been reported to cause intrauterine adhesions, surgical division of which has been associated with a 20% risk of placental adhesion (percreta or accreta) in a subsequent pregnancy (44). The standard approach to treatment of Asherman's syndrome is hysteroscopic surgical division and clearance of the adhesions, leaving something *in situ* using to keep the anterior and posterior uterine walls separated from each other (usually an inert intrauterine contraceptive device) and providing the patient with an oestrogen-based medication to stimulate uterine growth (45). Randomised trials to provide an evidence base for treatment are absent, and most clinicians see too few of these cases to be able to perform the operation and develop their own experiential data.

Uterine polyps are frequently found on ultrasound. Molecular markers of endometrial receptivity are reduced in the endometrium of women with endometrial polyps, which raises the possibility of reduction in embryo implantation (46). This might be an associative link, not causative, of course. Finding polyps during a treatment cycle for assisted conception can cause consternation amongst patients and clinicians, and patients will want to know if the polyps affect their chances of success. The evidence for an increase in fertility after polyp removal comes from nonblinded trials and observational studies, and therefore is poor quality, but most clinicians will still undertake the procedure. There has been only one RCT with 215 women, and this showed a higher likelihood of IUI-assisted pregnancy in the treatment group, RR 2.1 (95% CI 1.5–2.9) (47). In their comprehensive review on this topic, Alansari and Wardle discuss several case series, all with small numbers (6 and 8 patients), where pregnancies occurred in the treated group (48). This is another area where more research needs to be done, and management currently depends on the experience of the clinician involved.

6.4 PROXIMAL AND DISTAL INNER TUBAL DAMAGE

Chlamydia trachomatis is one of the most frequent causes of sexually transmitted disease (STD) in the United Kingdom and in the world; it accounts for about 45% of all tubal damage, both inner and outer tubal damage (49,50). It is a gram-negative bacterium, an 'obligate intracellular parasite', which means it cannot live outside a cell body. It has a delicate structure, but its ability to shift into elementary bodies with inclusion cysts enables it to survive. Other organisms beside Chlamydia can cause STDs, including gonorrhoea and mycoplasma, but Chlamydia is studied more because it is the most common. Chlamydia also leaves behind an antibody reaction to examine. Acute Chlamydia infection is less prevalent that one might think. Urine sampling for Chlamydia DNA was undertaken in a prospective cohort study (the ALSPAC study), where 51% of the 17-year-old cohort attended for sampling (51). Chlamydia prevalence was 1% in the 2,900 who agreed to Chlamydia screening. It was higher amongst the socially disadvantaged and found more in women than men (1.4% compared to 0.5%) (51). Those prevalence rates seem unexpectedly low and may be related to the self-selection of the 17 year olds attending and then consenting to the test.

The relationship between Chlamydia infection and its antibody reaction is well recognised (52). It appears this antibody reaction is long lasting, and while antibody levels reduce somewhat in the first 6 months after infection, it continues for years thereafter (53). Seventy-five percent of infertile women with damaged fallopian tubes have high titre levels, and the chances and severity of tubal damage are directly related to the level of antibody titre (54). The pertinent data are shown in Figure 6.3. Columns in black show the likelihood of any damage, hatched bar columns show the likelihood of serious damage; for example, at a titre of 1:1024, the likelihood of any damage is 80% and of serious damage over 40%. Women with antibody titres in the lower ranges were more likely to have pelvic infections than tubal damage. Titres in the upper ranges were more associated with tubal occlusion (54). The likelihood of ectopic pregnancy is elevated when the inner portion is damaged, and any pelvic infection in the past will lead to a risk of 10%–15%, with a doubling of risk with every subsequent infection. Ectopic pregnancy rate in healthy women is about 1%–2% (55), and an episode of pelvic infection increases that risk 10-fold (56).

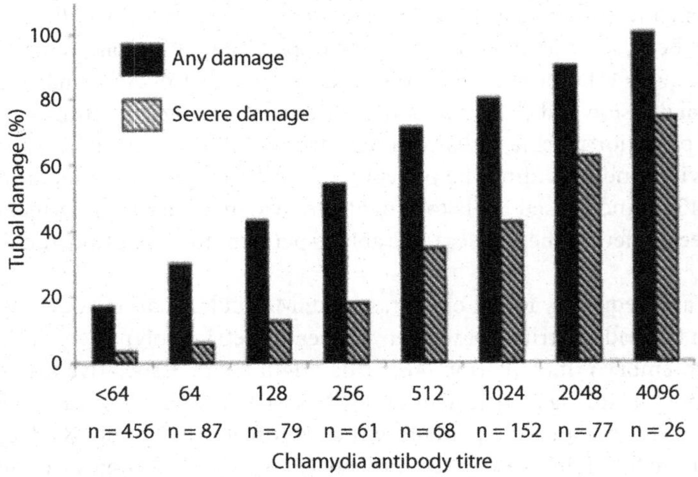

Figure 6.3 Frequency (%) of any tubal damage or severe tubal damage according to Chlamydia antibody titres. (From Akande VA et al. *Human Reproduction (Oxford, England)*. 2003;18(9):1841–7, with permission.)

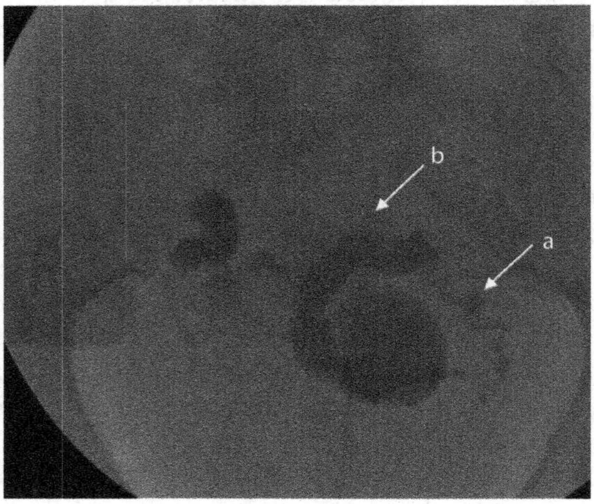

Figure 6.4 Radiology using hysterosalpingography demonstrating bilateral tubal occlusion, with (a) a long thread section of patent tube leading to (b) a dilated and distorted tube at the distal end. No spill is evident into the pelvic cavity.

The long-established investigations for fallopian tube patency are sonohysterography (see Section 3.5.1 in Chapter 3), hysterosalpingogram with a radio-opaque dye (see Figure 6.4) and laparoscopy combined with insufflation of the tubes with methylene blue dye (hydrochromopertubation). We have not adopted sonohysterography in Bristol for several reasons: cost streams, lack of confidence in the quality of the early images received and inertia due to the excellence of our hysterosalpingograpy service. Hysterosalpingography is an outpatient procedure, with some risk – and the dye can cause an allergic reaction, particularly if the woman is allergic to shellfish. This happens very infrequently (less than 1 in 500, I would estimate), particularly when a history of iodine or shellfish allergy is elicited. Allergic response can be covered by low-dose and short-term corticosteroids. Analgesic needs seldom exceed oral anti-inflammatory drugs. A sample operating procedure is presented in Figure 6.5.

Hysterosalpingogram

SETTING	Reproductive Medicine Clinic
FOR STAFF	Doctors & Trained nurses
PATIENTS	Patients undergoing infertility investigations

Woman's name

RMC patient ID

Topic	Initials	Date
At consultation 1. Clinical decision at time of consultation 2. If patient has raised antibodies to Chlamydia, check that both partners have had appropriate treatment (200 mg b.d ofloxacin for 10 days or Azithromycin 1G orally) 3. Fill out the request for the HSG on ICE 4. Prescription or over-the-counter medication to be taken on the day of HSG: Mefenamic acid 500 mg with food 2 h before and 4 h later *If any history of allergies (shellfish, iodine in particular):* 20 mg hydrocortisone orally 10 a.m. and 10 mg at 1 p.m. 5. Give patient the instruction sheet on HSG 6. Remind patient to abstain from sexual intercourse from the onset of the period to the day of the HSG		
When period starts, patient to ring xxxxxxxxxx to arrange appointments for the HSG (with the nurses). ***In nurse led clinic*** Nurses dispense the prescribed mefenamic acid and hydrocortisone. Nurses give patients self swab for NAATS. On Monday (two days) before planned HSG, check swab results. If positive for Chlamydia then both partners to be treated with Azithromycin 1G stat before HSG. **In X Ray department**, in a hospital HSG performed. Check patient has a review appointment. Advise patients to ring Gynaecology clinic if undue bleeding or pain.		

Audit Standards : Criteria	Target	Exceptions
Give the instruction sheet on HSG to the patient	100%	None
Prescription for analgesia and Antibiotics	100%	None

Figure 6.5 A standard operating procedure for a hysterosalpingogram.

The likelihood of ectopic pregnancy is elevated when the inner portion is damaged, and any pelvic infection in the past will lead to a risk of 10%–15%, with a doubling of risk with every subsequent infection. Ectopic pregnancy rate in healthy women is about 1%–2% (55), and an episode of pelvic infection increases that risk 10-fold (56).

Laparoscopy carries more risk – everything in the pelvis can be damaged: bowel (small and large intestine, usually transverse colon), bladder and blood vessels (usually one or other branch of the common iliac arteries). Guidelines to minimise the risk of damage produced jointly by the Royal College of Obstetricians and Gynaecologists, London (RCOG) and the British Society for Gynaecological Endoscopy advise that fluid be used to test that the Veress needle is in the peritoneal cavity and not in a hollow viscus (using the negative pressure of the cavity), and standardised methods for primary and secondary trocar entry (57). Using an open (Hassan) approach does not appear to reduce risks over the more traditional closed Veress needle approach (57). Nevertheless, if there is a history of conditions likely to cause adhesions, such as previous surgery, infection or inflammatory conditions, it is advisable to use an open laparoscopic approach (58). Dye insufflation rarely causes any immediate problems during surgery. Old infection could in theory be reactivated, so antibiotic cover is sensible.

Reassuringly, in women whose fallopian tubes are found to be normal at laparoscopy, natural conception rates are not affected whether titre levels are high or low (59). A very small study in the southwest of England indicated an increase in tubal damage when analysed by ethnicity for women of Afro-Caribbean background, but numbers were small and this is worthy of repetition in a bigger setting with a more diverse population (60).

One of the reasons Chlamydia as an organism causes so much damage relates to the way the body deals with its presence. As the infection is cleared, a protein (60-kDa heat shock protein – hsp60) is produced in large amounts. As the body's immune response clears this, the resultant inflammatory response is more destructive and damaging than would be expected, leading to more severe scar formation and tubal damage (61).

The two areas in the fallopian tube most prone to damage are the fimbrial end, because of the delicate fimbriae at the end of the tube, and the interstitial/cornual end, because this is the narrowest portion of the tube. Internal damage results from infection, but external damage (often from an STD, but do not forget the impact of inflammatory bowel disease, previous pelvic surgery and inflammation) can cause adhesion formation between the tube and other close structures, the ovary, the pelvic wall and the posterior aspect of the uterus.

Damage at the cornual and the fimbrial sites can be treated by surgical intervention. If the damage is proximal, the entry is made transcervically and the blockage cleared by selective salpingography, whereas an open or laparoscopic transabdominal approach is used for distal tube damage. As a technique, selective salpingography has been in use clinically since before 2000 (in fact, since the 1980s) and recanalisation success rates were reported of 80%. The ectopic pregnancy rate, in women whose tubes were made patent was 3% (62). Images for selective salpingography are shown in Figure 3.2. The findings of a 2012 Bristol study were in line with these results: we achieved tubal patency in at least one tube in 85% (84/99) of women and follow-up data were available for 96 women. A pregnancy rate of 24% (23/96) was achieved both in women treated for unilateral (27%) and bilateral (20%) tubal blockage. These pregnancies resulted from spontaneous natural conception or intrauterine insemination, not IVF. In women who conceived, the ectopic pregnancy rate was 13% (3/23) (63).

Distal tubal blockage, with occlusion of the fimbrial opening, is generally but not always demonstrated by the presence of a hydrosalpinx, a collection of fluid in the distal end of the tube (see the radiological image in Figure 6.4 and compare with Figure 3.2b where spill is seen). Hydrosalpinges have long been recognised as indicators of severe tubal damage and poor pregnancy outcomes with implantation rates at IVF (5.6%) significantly lower than the nonhydrosalpinx group (11.2%) (64). Twenty years ago, a Bristol group showed that draining hydrosalpinges prior to tubal surgery failed to improve the outcome for subsequent assisted conception, with implantation rates significantly reduced overall in the hydrosalpinx-drained group (8.0 versus 13.2% for controls; $p < 0.001$) (largely IVF) (65). They concluded that, 'Surgical drainage of distended hydrosalpinges appears to offer no benefit'. Subsequently, in a randomised controlled trial in 1999, the first high-quality data were published supporting salpingectomy for ultrasound evident hydrosalpinx in terms of improved implantation and live birth rates following IVF (66). They showed that live birth rates were 28.6% in the salpingectomy group and 16.3% in the control untreated group ($p = 0.045$), and in women with bilateral hydrosalpinges, the implantation rates were 25.6% versus 12.3% for the same groups ($p = 0.038$).

6.5 ADHESIONS AROUND THE TUBE AND OVARY

Peritubal adhesions are less of a problem than inner tubal lining damage – an illustration of what these look like can be seen in Figure 6.6. When extensive peritubal adhesions are evident, it is not

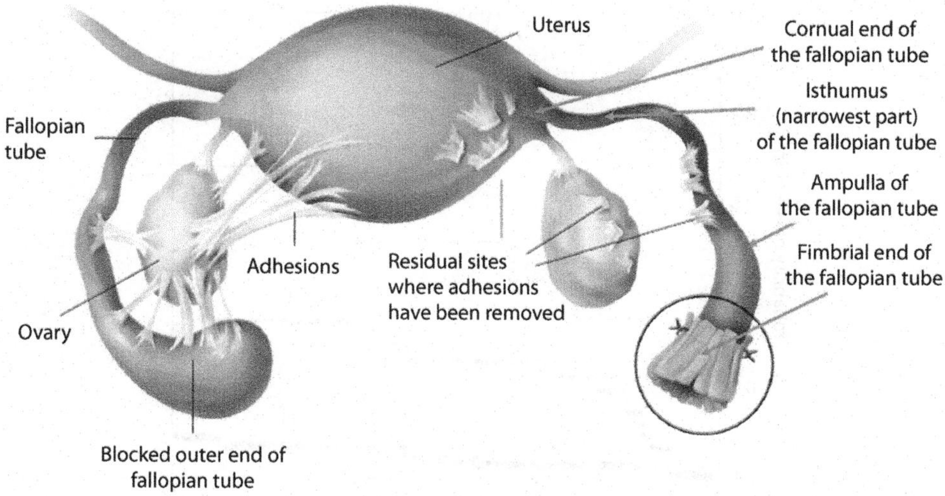

Figure 6.6 Adhesive complications can disrupt normal fallopian tube function. (From Wardle PG, Cahill DJ. 2005. *Understanding Infertility*: Family Doctor Publications, Poole, United Kingdom, with permission.)

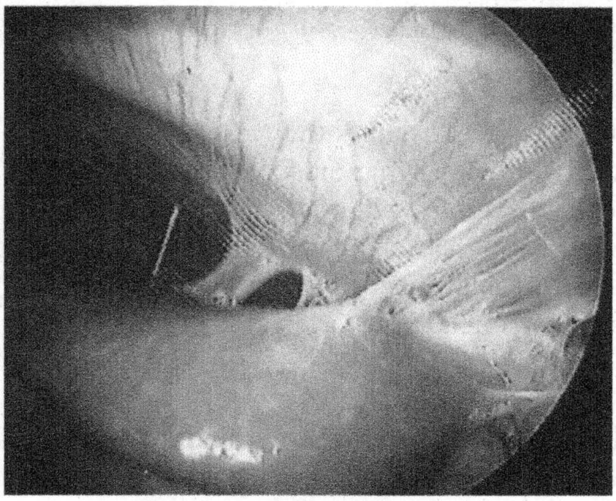

Figure 6.7 The presence of Fitz-Hugh Curtis syndrome generally indicating previous extensive Chlamydia infection.

uncommon to see evidence of Fitz-Hugh Curtis syndrome – fine 'violin string' adhesions between the liver and the diaphragm (occurs in up to 25% of cases [67]) – usually caused by Chlamydia (Figure 6.7).

Distal tubal damage (either peritubal and periovarian adhesions or hydrosalpinges) is best managed by abdominal surgery (open or laparoscopically). The most recent (2007) meta-analysis using studies published in the early 1990s, and employing intrauterine pregnancy rates to assess which modality is better found no difference between open or laparoscopic for mild/minor disease, but open surgery was better for severe disease (68). Worryingly, more and more fertility surgeons are turning to laparoscopic methods, and skills in open surgery are gradually being lost. Not surprisingly, intrauterine pregnancy rates in the severe tubal disease group were very low, irrespective of the surgical technique used. Data from Bristol also found that operating in the most severe group was not likely to result in live births – Figure 6.8 (69). This suggests these women would be better advised to go for IVF than undergoing tubal surgery. This paper used a locally derived classification system for tubal damage severity (70). Women should be counselled that ectopic pregnancy rates after treatment (IVF and surgery) are possibly as high as 10%–40%, respectively (71), and be

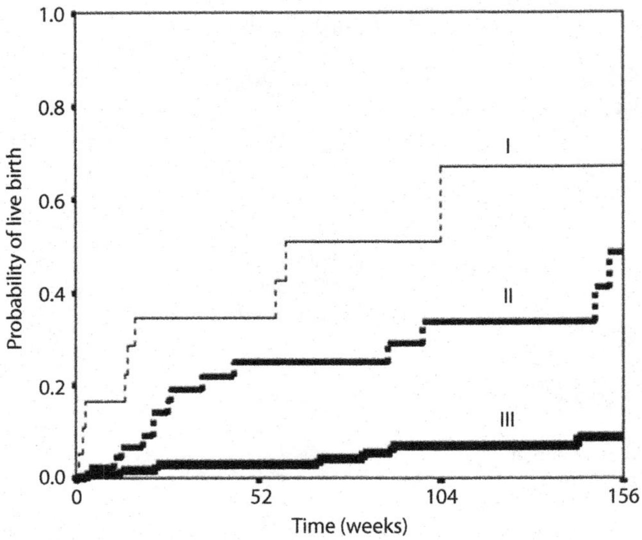

Figure 6.8 Kaplan–Meier curves showing probability of live birth following surgery in infertile women according to severity of tubal damage using the H&R classification (I = grade I; II = grade II; III = grade III) over a 3-year period. (From Akande VA et al. *BJOG: An International Journal of Obstetrics and Gynaecology*. 2004;111(11):1236–41, with permission.)

advised to present early for assessment should they miss a period and have a positive pregnancy test, even in the absence of any relevant symptoms.

When surgical approaches are being used, it is probably better to avoid using normal saline to flush tissue. As well as its pH being acidic, the peritoneal lining, when exposed to normal saline, appears to be predisposed to a greater propensity for adhesion formation (72), and alternatives such as Ringer's lactate may be better suited to the task (37).

6.6 CASE HISTORY

Jean is 23 and has been living with her current boyfriend for 14 months. She was not concerned about getting pregnant when they started to live together, and they have not used contraception. She had several long relationships in the past, and occasional one-night stands. Before meeting her current partner, she took the OCS Pill, as she didn't trust her partners enough to rely on them for contraception. Over the past 4 years, she has had several episodes of lower abdominal pain, some of which were thought to be grumbling appendicitis. She was admitted to hospital during the last episode because the pain was so severe. As part of the investigations, she had swabs taken from her cervix, some blood tests and a laparoscopy. The results indicated a diagnosis of Chlamydia infection, and an antibiotic was given to treat this. The laparoscopy showed that both her tubes were blocked at their outer (fimbrial) ends (see Figure 6.6). She was told that she was unlikely to conceive, and that surgery would be necessary if and when she wanted to get pregnant. It was explained that she should wait to have surgery until she wanted to conceive, as the best chance of success would be within the first year after surgery, see www.nice.org.uk/guidance/cg156 for details. Her partner was advised of his need for antibiotic treatment as well. They were both advised that, as they might possibly have other sexually transmitted diseases, they should visit the sexual health clinic. As yet, they have not decided to have any surgery.

6.6.1 Comments, investigations and management outlines

Having several partners and unprotected intercourse increases the risk of catching a sexually transmitted infection. Although the Pill is effective in preventing unwanted pregnancy, it is not of any

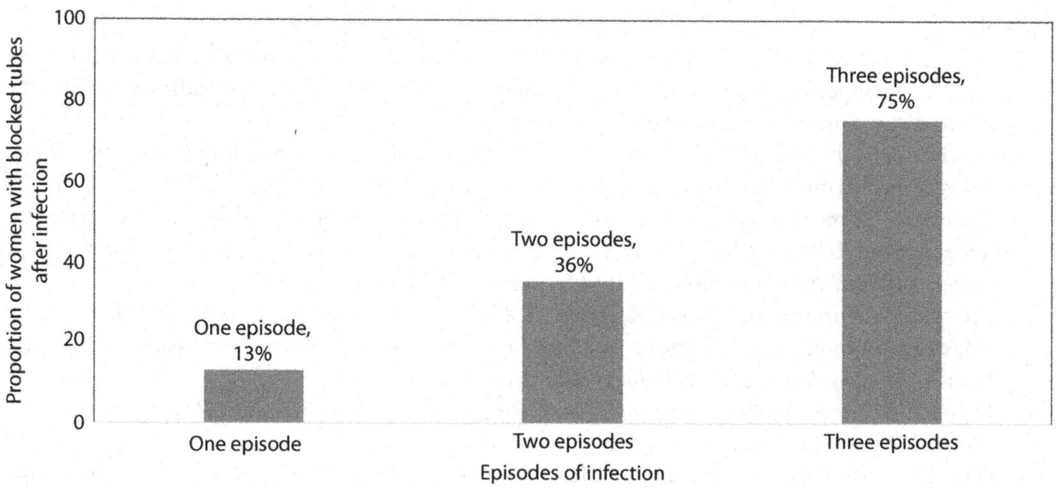

Figure 6.9 Likelihood of tubal blockage after one, two and three episodes of pelvic inflammatory disease. (Adapted from data in Westrom L. *American Journal of Obstetrics and Gynecology*. 1975; 121(5):707–13.)

value in the prevention of STIs. Education for young people needs to emphasise the need for barrier protection to protect against STIs and that a condom is the most effective. Condom use and altering sexual behaviour reduces the likelihood of contracting an STI (73). It seems likely that Jean's episodes of pain were in fact the result of pelvic inflammatory disease caused by Chlamydia. Each episode of infection has about a 30% chance of causing structural damage to the fallopian tubes (74) (see Figure 6.9 for actual proportions), with an increasing chance of tubal blockage with each episode of new infection. The cervical swabs did not grow bacteria – they often don't, as Chlamydia is particularly difficult to grow in the laboratory, and the swabs are designed instead to pick up Chlamydia DNA. The blood test looked at antibodies to Chlamydia infection, both recent and past, measuring IgM and IgG immunoglobulins. The particular immunoglobulin raised indicates recent or past infection, but evidence of past infection is what most research and clinical data refer to. In Jane's case, these results indicated considerable infection in the past and a current infection as well. Antibiotics such as ofloxacin or doxycycline are the recommended treatment and need to be taken by both partners. Providing there is no further damage from another infection, the chance of success, of getting pregnant within a year after surgery, is no better than 50%. A hysterosalpingogram would clarify the likelihood more accurately by giving information about the health of the inner tubal lining. Surgery should not be done until the couple are sure that they want to proceed with a pregnancy. The best chance of success is in the first year after surgery. If pregnancy hasn't occurred by then, it is unlikely to occur at all, usually because of damage to the delicate lining of the fallopian tubes. Repeat attempts at surgery are generally not worthwhile.

6.7 SUMMARY

The evidence for surgical management of congenital uterine anomalies is limited and not supportive of intervention. Uterine fibroid surgery can cause significant blood loss and adhesion formation, but careful attention to haemostasis intraoperatively and use of Ringer's lactate can address this satisfactorily. Peritubal adhesions and blockage are usually caused by Chlamydia. Careful surgical technique in both laparoscopic and open surgery, paying attention to irrigation fluids used and keeping the surgical field blood free and nontouch, will provide best results for tubal surgery. Evidence quality for tubal surgery is good; for fibroid surgery, it is generally of poor quality.

REFERENCES

1. Venetis CA, Papadopoulos SP, Campo R, Gordts S, Tarlatzis BC, Grimbizis GF. Clinical implications of congenital uterine anomalies: A meta-analysis of comparative studies. *Reproductive Biomedicine Online*. 2014;29(6):665–83.
2. Dunn CL, Kelly RW, Critchley HO. Decidualization of the human endometrial stromal cell: An enigmatic transformation. *Reproductive Biomedicine Online*. 2003;7(2):151–61.
3. Kupesic S. Clinical implications of sonographic detection of uterine anomalies for reproductive outcome. *Ultrasound in Obstetrics & Gynecology: the Official Journal of the International Society of Ultrasound in Obstetrics and Gynecology*. 2001;18(4):387–400.
4. Crocker IP, Wareing M, Ferris GR, Jones CJ, Cartwright JE, Baker PN et al. The effect of vascular origin, oxygen, and tumour necrosis factor alpha on trophoblast invasion of maternal arteries *in vitro*. *Journal of Pathology*. 2005;206(4):476–85.
5. Tanos V, Berry KE. Benign and malignant pathology of the uterus. *Best Practice & Research Clinical Obstetrics & Gynaecology*. 2018;46:12–30.
6. Wise LA, Laughlin-Tommaso SK. Epidemiology of uterine fibroids: From menarche to menopause. *Clinical Obstetrics and Gynecology*. 2016;59(1):2–24.
7. Cramer SF, Patel A. The frequency of uterine leiomyomas. *American Journal of Clinical Pathology*. 1990;94(4):435–8.
8. Reis FM, Bloise E, Ortiga-Carvalho TM. Hormones and pathogenesis of uterine fibroids. *Best Practice & Research Clinical Obstetrics & Gynaecology*. 2016;34:13–24.
9. Laughlin SK, Schroeder JC, Baird DD. New directions in the epidemiology of uterine fibroids. *Seminars in Reproductive Medicine*. 2010;28(3):204–17.
10. Klatsky PC, Tran ND, Caughey AB, Fujimoto VY. Fibroids and reproductive outcomes: A systematic literature review from conception to delivery. *American Journal of Obstetrics and Gynecology*. 2008;198(4):357–66.
11. Pritts EA, Parker WH, Olive DL. Fibroids and infertility: An updated systematic review of the evidence. *Fertility and Sterility*. 2009;91(4):1215–23.
12. Sunkara SK, Khairy M, El-Toukhy T, Khalaf Y, Coomarasamy A. The effect of intramural fibroids without uterine cavity involvement on the outcome of IVF treatment: A systematic review and meta-analysis. *Human Reproduction (Oxford, England)*. 2010;25(2):418–29.
13. Bernard G, Darai E, Poncelet C, Benifla JL, Madelenat P. Fertility after hysteroscopic myomectomy: Effect of intramural myomas associated. *European Journal of Obstetrics, Gynecology, and Reproductive Biology*. 2000;88(1):85–90.
14. Bosteels J, Kasius J, Weyers S, Broekmans FJ, Mol BW, D'Hooghe TM. Hysteroscopy for treating subfertility associated with suspected major uterine cavity abnormalities. *Cochrane Database of Systematic Reviews*. 2015:2;Cd009461.
15. Metwally M, Cheong YC, Horne AW. Surgical treatment of fibroids for subfertility. *Cochrane Database of Systematic Reviews*. 2012;11:Cd003857.
16. Lethaby A, Vollenhoven B, Sowter M. Pre-operative GnRH analogue therapy before hysterectomy or myomectomy for uterine fibroids. *Cochrane Database of Systematic Reviews*. 2000:2;Cd000547.
17. Genazzani AR, Pluchino N, Bernardi F, Centofanti M, Luisi M. Beneficial effect of tibolone on mood, cognition, well-being, and sexuality in menopausal women. *Neuropsychiatric Disease and Treatment*. 2006;2(3):299–307.
18. Moroni RM, Martins WP, Ferriani RA, Vieira CS, Nastri CO, Candido Dos Reis FJ et al. Add-back therapy with GnRH analogues for uterine fibroids. *Cochrane Database of Systematic Reviews*. 2015:3:Cd010854.
19. Woodhead N, Pounds R, Irani S, Pradhan P. Ulipristal acetate for uterine fibroids: 2 years of real world experience in a UK hospital. *Journal of Obstetrics and Gynaecology: Journal of the Institute of Obstetrics and Gynaecology*. 2018:29;1–5.

20. Donnez J, Vazquez F, Tomaszewski J, Nouri K, Bouchard P, Fauser BC et al. Long-term treatment of uterine fibroids with ulipristal acetate. *Fertility and Sterility*. 2014;101(6):1565–73. el–18.
21. European Medicines Agency. *Esmya: New Measures to Minimise Risk of Rare but Serious Liver Injury*. London: European Medicines Agency; 2018. [Professional advise on ullipristal use in women with fibroids.] Available from: https://www.ema.europa.eu/en/medicines/human/referrals/esmya.
22. Liang E, Brown B, Rachinsky M. A clinical audit on the efficacy and safety of uterine artery embolisation for symptomatic adenomyosis: Results in 117 women. *Australian & New Zealand Journal of Obstetrics & Gynaecology*. 2018;58(4):454–9.
23. Liang E, Brown B, Kirsop R, Stewart P, Stuart A. Efficacy of uterine artery embolisation for treatment of symptomatic fibroids and adenomyosis - an interim report on an Australian experience. *Australian & New Zealand Journal of Obstetrics & Gynaecology*. 2012;52(2):106–12.
24. Gupta JK, Sinha A, Lumsden MA, Hickey M. Uterine artery embolization for symptomatic uterine fibroids. *Cochrane Database of Systematic Reviews*. 2014:12;Cd005073.
25. Hirst A, Dutton S, Wu O, Briggs A, Edwards C, Waldenmaier L et al. A multi-centre retrospective cohort study comparing the efficacy, safety and cost-effectiveness of hysterectomy and uterine artery embolisation for the treatment of symptomatic uterine fibroids. The HOPEFUL study. *Health Technology Assessment (Winchester, England)*. 2008;12(5):1–248, iii.
26. Torre A, Fauconnier A, Kahn V, Limot O, Bussierres L, Pelage JP. Fertility after uterine artery embolization for symptomatic multiple fibroids with no other infertility factors. *European Radiology*. 2017;27(7):2850–9.
27. Royal College of Obstetricians and Gynaecologists. *Uterine Artery Embolisation in the Management of Fibroids*. London: RCOG; 2013.
28. Xia EL, Duan H, Huang XW, Zheng J, Yu D. Transcervical resection of myoma in treatment of hysteromyoma, experience in 962 x\cases. *Zhonghua yi xue za zhi*. 2005;85(3):173–6.
29. Cravello L, Agostini A, Beerli M, Roger V, Bretelle F, Blanc B. Results of hysteroscopic myomectomy. *Gynecologie, Obstetrique & Fertilite*. 2004;32(9):825–8.
30. Sirkeci RF, Belli AM, Manyonda IT. Treating symptomatic uterine fibroids with myomectomy: Current practice and views of UK consultants. *Gynecological Surgery*. 2017;14(1):11.
31. Stanhiser J, Mouille B, Flyckt R, Goldberg J, Falcone T, Goodman LR. Trends over time and surgical outcomes of abdominal, mini-laparotomy, and traditional and robotic-assisted laparoscopy with and without tandem mini-laparotomy: A comparison of myomectomy techniques. *Journal of Minimally Invasive Gynecology*. 2015;22(6s):S1.
32. Glasser MH. Minilaparotomy: A minimally invasive alternative for major gynecologic abdominal surgery. *The Permanente Journal*. 2005;9(1):41–5.
33. Glaser LM, Friedman J, Tsai S, Chaudhari A, Milad M. Laparoscopic myomectomy and morcellation: A review of techniques, outcomes, and practice guidelines. *Best Practice & Research Clinical Obstetrics & Gynaecology*. 2018;46:99–112.
34. Parker WH, Pritts EA, Olive DL. What is the future of open intraperitoneal power-morcellation of fibroids? *Clinical Obstetrics and Gynecology*. 2016;59(1):73–84.
35. Bean EM, Cutner A, Holland T, Vashisht A, Jurkovic D, Saridogan E. Laparoscopic myomectomy: A single-center retrospective review of 514 patients. *Journal of Minimally Invasive Gynecology*. 2017;24(3):485–93.
36. Dedes IA, Squires RM, Thomas EK, Mills MS, Cahill DJ. Should we still be undertaking open myomectomies? A five-year retrospective case review. *Gynaecology and Perinatology* 2016;1(1):9–17.
37. Sahakian V, Rogers RG, Halme J, Hulka J. Effects of carbon dioxide-saturated normal saline and Ringer's lactate on postsurgical adhesion formation in the rabbit. *Obstetrics and Gynecology*. 1993;82(5):851–3.

38. Subramanian S, Clark MA, Isaacson K. Outcome and resource use associated with myomectomy. *Obstetrics and Gynecology*. 2001;98(4):583–7.
39. Vannuccini S, Tosti C, Carmona F, Huang SJ, Chapron C, Guo S-W et al. Pathogenesis of adenomyosis: An update on molecular mechanisms. *Reproductive Biomedicine Online*. 2017;35(5):592–601.
40. de Souza NM, Brosens JJ, Schwieso JE, Paraschos T, Winston RM. The potential value of magnetic resonance imaging in infertility. *Clinical Radiology*. 1995;50(2):75–9.
41. Devlieger R, D'Hooghe T, Timmerman D. Uterine adenomyosis in the infertility clinic. *Human Reproduction Update*. 2003;9(2):139–47.
42. Rocha TP, Andres MP, Borrelli GM, Abrao MS. Fertility-sparing treatment of adenomyosis in patients with infertility: A systematic review of current options. *Reproductive Sciences (Thousand Oaks, Calif)*. 2018;25(4):480–6.
43. Miller-Fellows SC, Howard L, Kramer R, Hildebrand V, Furin J, Mutuku FM et al. Cross-sectional interview study of fertility, pregnancy, and urogenital schistosomiasis in coastal Kenya: Documented treatment in childhood is associated with reduced odds of subfertility among adult women. *PLoS Neglected Tropical Diseases*. 2017;11(11):e0006101.
44. Ibrahim MI, Raafat TA, Ellaithy MI, Aly RT. Risk of postpartum uterine synechiae following uterine compression suturing during postpartum haemorrhage. *Australian & New Zealand Journal of Obstetrics & Gynaecology*. 2013;53(1):37–45.
45. Capella-Allouc S, Morsad F, Rongières-Bertrand C, Taylor S, Fernandez H. Hysteroscopic treatment of severe Asherman's syndrome and subsequent fertility. *Human Reproduction*. 1999;14(5):1230–3.
46. Rackow BW, Jorgensen E, Taylor HS. Endometrial polyps affect uterine receptivity. *Fertility and Sterility*. 2011;95(8):2690–2.
47. Perez-Medina T, Bajo-Arenas J, Salazar F, Redondo T, Sanfrutos L, Alvarez P et al. Endometrial polyps and their implication in the pregnancy rates of patients undergoing intrauterine insemination: A prospective, randomized study. *Human Reproduction (Oxford, England)*. 2005;20(6):1632–5.
48. Alansari LM, Wardle P. Endometrial polyps and subfertility. *Human Fertility (Cambridge, England)*. 2012;15(3):129–33.
49. Price MJ, Ades AE, Welton NJ, Macleod J, Turner K, Simms I et al. How much tubal factor infertility is caused by Chlamydia? Estimates based on serological evidence corrected for sensitivity and specificity. *Sexually Transmitted Diseases*. 2012;39(8):608–13.
50. Menon S, Timms P, Allan JA, Alexander K, Rombauts L, Horner P et al. Human and pathogen factors associated with *Chlamydia trachomatis*-related infertility in women. *Clinical Microbiology Reviews*. 2015;28(4):969–85.
51. Crichton J, Hickman M, Campbell R, Heron J, Horner P, Macleod J. Prevalence of Chlamydia in young adulthood and association with life course socioeconomic position: Birth cohort study. *PLOS ONE*. 2014;9(8):e104943.
52. Conway D, Glazener CM, Caul EO, Hodgson J, Hull MG, Clarke SK et al. Chlamydial serology in fertile and infertile women. *Lancet (London, England)*. 1984;1(8370):191–3.
53. Horner PJ, Wills GS, Reynolds R, Johnson AM, Muir DA, Winston A et al. Effect of time since exposure to *Chlamydia trachomatis* on Chlamydia antibody detection in women: A cross-sectional study. *Sexually Transmitted Infections*. 2013;89(5):398–403.
54. Akande VA, Hunt LP, Cahill DJ, Caul EO, Ford WC, Jenkins JM. Tubal damage in infertile women: Prediction using Chlamydia serology. *Human Reproduction (Oxford, England)*. 2003;18(9):1841–7.
55. Barash JH, Buchanan EM, Hillson C. Diagnosis and management of ectopic pregnancy. *American Family Physician*. 2014;90(1):34–40.
56. Hull MG, Cahill DJ. Female infertility. *Endocrinology and Metabolism Clinics of North America*. 1998;27(4):851–76.

57. Sutton CJG, Philips K. *Preventing Entry-Related Gynaecological Laparoscopic Injuries*. RCOG; 2008.
58. Hasson HM. A modified instrument and method for laparoscopy. *American Journal of Obstetrics and Gynecology*. 1971;110(6):886–7.
59. El Hakim EA, Epee M, Draycott T, Gordon UD, Akande VA. Significance of positive Chlamydia serology in women with normal-looking fallopian tubes. *Reproductive Biomedicine Online*. 2009;19(6):847–51.
60. Oworu O, Jones K, Moffatt J. *Is Ethnicity a Factor in Patients Receiving Assisted Reproductive Treatments?* Human Fertility: Cambridge, England, Taylor & Francis. 2016, p. 63–9.
61. Witkin SS, Minis E, Athanasiou A, Leizer J, Linhares IM. *Chlamydia trachomatis*: The Persistent Pathogen. *Clinical and Vaccine Immunology: CVI*. 2017;24(10).
62. Thurmond AS, Machan LS, Maubon AJ, Rouanet JP, Hovsepian DM, Moore A et al. A review of selective salpingography and fallopian tube catheterization. *Radiographics: A Review Publication of the Radiological Society of North America, Inc*. 2000;20(6):1759–68.
63. Hussain M, Cahill D. Tubal subfertility: Role of selective salpingography and transcervical tubal catherization. *Annual British Fertility Society Meeting*; January 2012; Leeds: British Fertility Society 2012.
64. Fleming C, Hull MG. Impaired implantation after *in vitro* fertilisation treatment associated with hydrosalpinx. *British Journal of Obstetrics and Gynaecology*. 1996;103(3):268–72.
65. Sowter MC, Akande VA, Williams JA, Hull MG. Is the outcome of *in-vitro* fertilization and embryo transfer treatment improved by spontaneous or surgical drainage of a hydrosalpinx? *Human Reproduction (Oxford, England)*. 1997;12(10):2147–50.
66. Strandell A, Lindhard A, Waldenstrom U, Thorburn J, Janson PO, Hamberger L. Hydrosalpinx and IVF outcome: A prospective, randomized multicentre trial in Scandinavia on salpingectomy prior to IVF. *Human Reproduction (Oxford, England)*. 1999;14(11):2762–9.
67. Peter NG, Clark LR, Jaeger JR. Fitz-Hugh Curtis syndrome: A diagnosis to consider in women with right upper quadrant pain. *Cleveland Clinic Journal of Medicine*. 2004;71(3):233–9.
68. Ahmad G, Watson AJ, Metwally M. Laparoscopy or laparotomy for distal tubal surgery? A meta-analysis. *Human Fertility (Cambridge, England)*. 2007;10(1):43–7.
69. Akande VA, Cahill DJ, Wardle PG, Rutherford AJ, Jenkins JM. The predictive value of the 'Hull & Rutherford' classification for tubal damage. *BJOG: An International Journal of Obstetrics and Gynaecology*. 2004;111(11):1236–41.
70. Rutherford AJ, Jenkins JM. Hull and Rutherford classification of infertility. *Human Fertility (Cambridge, England)*. 2002;5(1 Suppl):S41–5.
71. Schippert C, Soergel P, Staboulidou I, Bassler C, Gagalick S, Hillemanns P et al. The risk of ectopic pregnancy following tubal reconstructive microsurgery and assisted reproductive technology procedures. *Archives of Gynecology and Obstetrics*. 2012;285(3):863–71.
72. Cwalinski J, Staniszewski R, Baum E, Jasinski T, Mackowiak B, Breborowicz A. Normal saline may promote formation of peritoneal adhesions. *International Journal of Clinical and Experimental Medicine*. 2015;8(6):8828–34.
73. Fernane, S, Fruitman, H, Topping, T, Clarke, D, Lane, A, Secan, V. Effective methods to prevent sexually transmitted infections. In: *Region of Peel Public Health,*. Peel, ONT. 2012. pp. 53.
74. Westrom L. Effect of acute pelvic inflammatory disease on fertility. *American Journal of Obstetrics and Gynecology*. 1975;121(5):707–13.

What to do if nothing wrong can be found and how to answer when a couple asks 'what can we do to improve our fertility?'

7

OVERVIEW

This chapter explores a common challenge in infertility medicine – how to manage a couple in whom all the tests are normal and where no cause can be found for their subfertility. It covers endometriosis, alternative medicine and lifestyle and environmental factors.

It also provides information for clinicians to use when advising couples about lifestyle changes they can make to improve the possibility of natural conception, as well as addressing those areas which are not amenable to change.

7.1 MAKING THE DIAGNOSIS OF UNEXPLAINED INFERTILITY

For almost a third of couples attending specialist clinics, the results of all their investigations will be normal. This situation is called unexplained infertility, a term which can be erroneously interpreted by patients to mean 'not fully investigated infertility', and which can engender frustration and despair. The label or diagnosis of unexplained infertility has strict definition criteria (1,2). These include:

- The menstrual cycle of the woman is normal (3–6 weeks in length)
- Seminal analysis of the man yields normal results
- The postcoital sperm penetration of mucus (if done) is normal (most units will not do this test – see later text)
- Coital activity has been at least twice a week, for at least 1 year (1)
- Laparoscopic findings are normal OR the woman had previously been pregnant, when the pelvic and tubal state can be assumed to be normal (2)

A randomised controlled trial on the effectiveness of the postcoital test in infertility investigations did not find it to be helpful (3), and consequently other authors and clinicians have removed the postcoital test from the investigations required to assess unexplained infertility (4). In the latter review, they considered that key investigations to enable the diagnosis of unexplained infertility to be made were: semen analysis, assessment of ovulation and evaluation of tubal patency by hysterosalpingogram or laparoscopy (as above) (4).

Unexplained infertility comprises a sizable proportion of cases presenting to reproductive medicine clinics. Estimates range from 15%–30%. Table 7.1 illustrates this point. (Note that in about 16% of cases, infertility has more than one cause, for example, tubal damage *and* ovulatory problems, or tubal damage with a sperm disorder, which is why in Table 7.1, the percentage total of all causes of infertility comes to 116.) Table 7.1 mentions endometriosis scarring as a potential cause of infertility, but it is by no means generally accepted that endometriosis has such a role. Debate on the relationship between endometriosis and infertility is in some respects akin to the debate about fibroids and fertility. A description of endometriosis is given in Section 7.1.1.

Table 7.1 The common causes of infertility (the total exceeds 100% because 16% of couples have more than one cause for their infertility).

Cause		Percentage
Women		
	Ovulation failure	20
	Fallopian tube damage	15
	Scarring from endometriosis	6
	Cervical mucus problems	3
Men		
	Poor sperm function/low sperm count	25
	Failure of sperm production/release	2
Both partners		
	Unexplained infertility	28
	Problems with intercourse	6
	Miscellaneous other causes	11
Total		116

Source: From Wardle PG, Cahill DJ. *Understanding Infertility*: Family Doctor Publications, Dorset United Kingdom, 2005, with permission (93).

Reviewing the literature on unexplained infertility, there are many areas under investigation as potential causes of unexplained infertility. These include the impact of heavy metals on the endometrium, the role of thyroid function and vitamin D deficiency, autoimmune conditions, food intolerance – and many others (26,27). It is right that couples should be fully investigated, but a point must come when investigations stop and a treatment plan is initiated to enable them to achieve their goal of a child. Sadeghi, writing in 2015, recognised this and considered the problem related to the disagreement between clinicians as to what constitutes 'adequate investigation' and 'unexplained infertility' (28). How many colleagues do we know who favour tests and investigations without which they believe the patient has not been fully investigated? I struggle to see the value of obscure diagnoses such as tubal peristaltic dysfunction (29) for a couple unless it provides direction to an effective treatment. I read sentiments like these almost 30 years ago in an editorial by Patrick Taylor titled 'When Is Enough Enough?' (30). He advocated gathering enough information within a reasonable amount of time to provide enough knowledge for the couple and their doctor to decide what to do.

7.1.1 Endometriosis

Endometriosis is a condition characterised by dysmenorrhoea (painful periods), dyspareunia (painful intercourse) and perimenstrual pain (pain around the time of periods, classically just before and after periods) and less frequently dyschezia (painful defaecation). The condition is caused by endometrial tissue which, for reasons unknown, becomes located in the peritoneal cavity or other sites. For the diagnosis to be made, histology must confirm endometrial glandular and stromal tissue, but most clinicians make the diagnosis without histological confirmation. Endometriosis in the peritoneal cavity usually expresses itself as minor lesions on the peritoneal surface or by more major lesions in the ovary, and causes adhesions between the ovary and surrounding structures such as – the pelvic side wall, the uterus or the fallopian tube (see the American Society for Reproductive Medicine's paper for the classification details) (5).

Some argue that any perceived association between endometriosis and fertility is only because it is more common in women who have not been pregnant for some time (6). In a large population study, endometriosis, both minor and major, was found in 6% of infertile couples (2). Bristol reproductive services took an especial interest in endometriosis and its effects on infertility and published several papers. It was a consistent finding that endometriosis had a negative effect on oocyte fertilisability and on the likelihood of conception (7–10). Most of the work focussed on the details of fertilisation rates of oocytes, but Figure 7.1 shows distinct differences in population data over time between women with unexplained infertility and women with only minor endometriosis whereby those with endometriosis had significantly lower pregnancy and live birth rates (10). The Bristol view of the impact of endometriosis is shared by many others and explored in detail in a review by D'Hooghe et al. (11). D'Hooghe makes potent and cogent arguments supporting the theory that endometriosis and reduced fertility are causally linked. While medical treatment is recognised not to have any impact on the fertility associated with endometriosis (12), a review of 10 RCTs found moderate-quality evidence that surgical management of endometriosis had beneficial effect. Three RCTs, $n = 528$, achieved an increased clinical pregnancy rate, OR 1.89. CI 1.25–2.86, $p = 0.003$, while 2 RCTs, $n = 382$, found increased ongoing pregnancy rates and live birth rates with OR 1.94%, 95% CI 1.20–3.16, $p = 0.007$ (13).

In studies with detailed endocrinology and fertilisation and implantation rates, there are conflicting data. The issue is complicated by different opinions about whether natural cycles with no exogenous gonadotrophins (6,14) or cycles augmented with gonadotrophins (supported by pragmatists) (7,8) are included in trials. Table 7.2 illustrates this; Mahmood and Cahill are both in natural cycles and have quite different findings. There are studies which point to a reduction in fertilisation rates in women with minor endometriosis (14) or to no difference (6,15). Some studies suggest endocrine and follicular dynamics dysfunction, although many were underpowered and therefore inconclusive. In their papers, Bancroft et al. and Tummon et al. both pointed to abnormal endocrinology with LH surge dysfunction, lower oestradiol and LH levels at the LH surge (16,17).

Population-based studies on pregnancy rates over time, between women with endometriosis compared to controls, provide another perspective on this question. Again, there is disparity between studies. The Canadian Collaborative Group on Endometriosis followed 331 women for 9 months. They found no difference in fecundity rates (in this case, getting pregnant and carrying a pregnancy for more than 20 weeks gestation) between women with endometriosis and controls (2.5 versus 3.5 per 100 person-months) (18). Data from one Australian paper, where women were followed for 6 months, and in which the use of donor sperm was controlled for sperm disorders, women with endometriosis had significantly lower cumulative conception rates compared

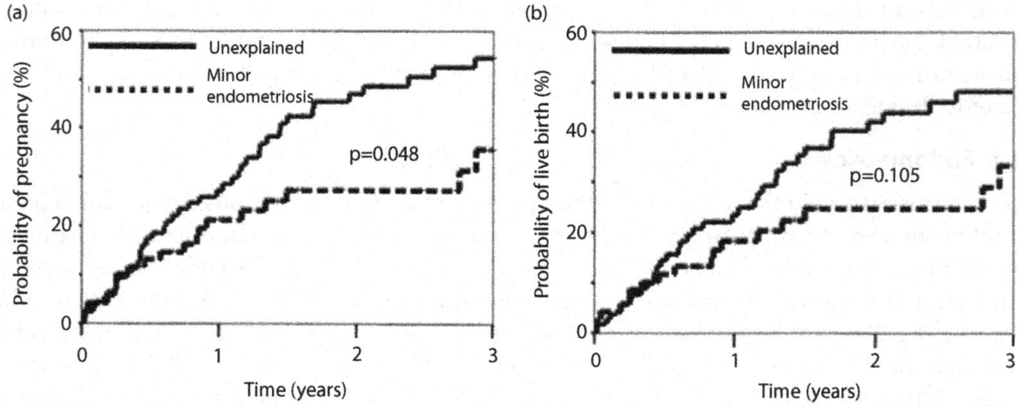

Figure 7.1 Kaplan-Meier curves showing differences and estimated probabilities of (a) pregnancy and (b) live birth for infertile women with minor endometriosis and unexplained infertility. The P-value refers to the overall difference among the curves using the log rank test. (From Akande VA et al. *Human Reproduction (Oxford, England)*. 2004;19(1):96–103, with permission.)

Table 7.2 Fertilisation rates in studies comparing women with endometriosis with controls.

			Fertilisation rate		Implantation rate		
Author	**n**	**Year**	**Endometriosis**	**Controls**	**Endometriosis**	**Controls**	**Significance**
Wardle	81	1985	57	27	–	–	<0.001
Mahmood	27	1991	53	63	–	–	NS
Mills	1577	1992	48	69	–	–	<0.001
Mills	1577	1992			10	12	NS
Fleming	2701	1994	71	61			<0.001
Cahill	129	1997	46	69	–	–	<0.01
Hull	11,076	1998	60	56	–	–	<0.001
Hull	11,076	1998			11	15	NS
Omland	752	2005	68	69	58	60	NS
Gonzalez-Comadran	22,500	2017	62	60	–	–	NS

Source: Adapted from data in Mahmood TA et al. *British Journal of Obstetrics and Gynaecology*. 1991;98(6):573–7; Wardle PG et al. *Lancet (London, England)*. 1985;2(8449):236–9; Mills MS et al. *Human Reproduction (Oxford, England)*. 1992;7(4):490–4; Cahill DJ et al. *Journal of Assisted Reproduction and Genetics*. 1997;14(10):554–7; Omland AK et al. *Human Reproduction (Oxford, England)*. 2005;20(3):722–7; Gonzalez-Comadran M et al. *Reproductive Biology and Endocrinology: RB&E*. 2017;15(1):8; Fleming CF et al. *Hum Reprod*. 1994;9(Suppl. 1):117; Hull MG et al. *BMJ (Clinical Research Ed)*. 1992;304(6840):1465–9; Hull MG et al. *Human Reproduction (Oxford, England)*. 1998;13(7):1825–30.

to controls (3% vs 12%, respectively) (19). However, Table 7.1 shows cumulative conception data in 192 women followed for 3 years; pregnancy rates (but not live birth rates) were lower in women with endometriosis compared to controls (10).

D'Hooghe's review (11) would be a good place to finish except the story continues and a very large controlled study compared fertilisation rates in simulated cycles included 22,500 women, 3500 with endometriosis and 18,000 controls (20). They found no differences in fertilisation rate between controls and women with endometriosis, 61.24% vs 60.41%. A summary of several papers which examined fertilisation rates in women with endometriosis compared to controls is presented in Table 7.2. There are a similar number of papers supporting the negative impact of endometriosis as that refute it. Consequently, the definitive answer to the question 'Does endometriosis cause infertility?' still eludes us.

A recurring theme in this book is that clinicians jump to treating conditions before there is clarity about whether that condition has any effect on fertility. A Cochrane review examined all the publications available on whether surgery for endometriosis has any effect on fertility (13). Two principal studies are included in this review – one from the Canadian Collaborative group (24) referred to earlier, the other from the Italian Endometriosis Study group (25). The Canadian group's paper is the biggest contributor to the data. The outcome of the review is that laparoscopic surgery with ablation of endometriotic lesions, when compared with diagnostic laparoscopy only, was associated with an increased clinical pregnancy rate (OR 1.89, 95% CI 1.25 to 2.86). Cochrane made it clear that the judgement was based on moderate-quality evidence (13). It seems reasonable, therefore, to offer laparoscopy and ablation of endometriosis when it is found, and to obtain informed consent for ablation in addition, from women undergoing a diagnostic laparoscopy for infertility.

7.2 MANAGEMENT AND TREATMENT OF UNEXPLAINED INFERTILITY

Expectant management is a deliberate policy of nonintervention (combined with information and support) which is chosen in any condition, because it has some virtue – delaying otherwise unnecessary investigations or treatment (32). The approach to unexplained infertility advised by the NICE guidelines is withholding investigations until 2 years of trying have passed, based on the knowledge that 95% of couples will conceive in that time (31).

Couples who have had unexplained infertility for less than 3 years have a normal chance of conception if they keep trying. The first phase of the management of unexplained fertility is termed expectant waiting, when the couple is encouraged to continue trying to conceive naturally for a total of 3 years. Once they have been trying for more than 3 years, or if the woman is over 35 years of age, they need to consider more active fertility treatments, such as IUI or IVF. The evidence for this is presented in Figure 7.2 (10). If the couple has not conceived after trying for 3 years, the chances of their being successful if they carry on are of the order of 1%–2% per month.

Ovulation induction agents (clomifene – orally – or gonadotrophins – by injection) improve the chances of pregnancy by stimulating the release of more than one egg each month. Clomifene on its own is generally recognised not to improve the chance of pregnancy in women with unexplained infertility (33). Intrauterine insemination in combination with ovulation induction using clomifene or gonadotrophins gives a higher chance of pregnancy (15%–20%/cycle). IUI is only suitable if the woman has healthy fallopian tubes and if sperm numbers and sperm function tests are normal. Data from elsewhere are even more encouraging about the use of IUI, a relatively simple technique; with women who were treated with IUI, 35% delivered after this – with reported multiple pregnancy rates of 12% (34).

These are (relatively) simple treatments for couples with unexplained infertility. They are not trying to overcome any particularly defined problem, simply increasing the chance of conception. In the same way, more advanced treatments such as assisted conception are useful and effective methods of achieving a pregnancy in unexplained infertility, in the right circumstances. Indeed, it might be wise to resort more quickly to assisted conception if a couple have been trying for a long time or the woman is nearing the age of 40, recognising the negative impact of age on the likelihood of conception.

Even if couples undergo IVF treatment and don't conceive, they still have the potential to conceive spontaneously. We found a surprisingly high likelihood of spontaneous conception, particularly for clinical situations in which the infertility was unexplained (35). Figure 7.3 illustrates this. Couples with unexplained infertility had a 35% chance of conceiving in the 3 years following an unsuccessful cycle of IVF treatment. A similar paper published from the Danish Assisted Reproduction Technology (ART) register noted encouraging pregnancy rates. They examined couples who were

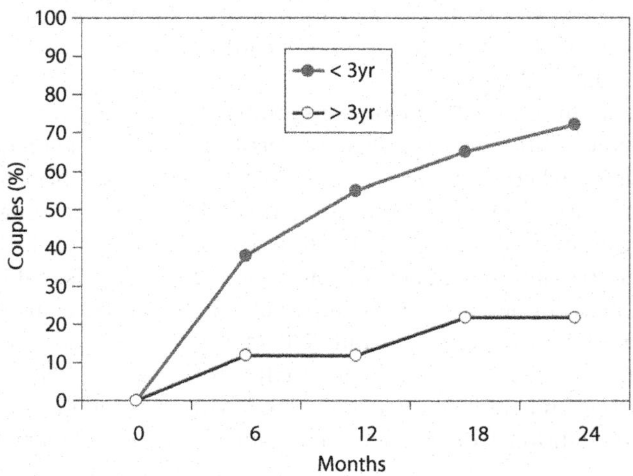

Figure 7.2 Conception rates in couples with unexplained infertility, with less than and greater than 3 years infertility. (From Hull MG, Cahill DJ. *Endocrinology and Metabolism Clinics of North America*. 1998;27(4):851–76, with permission.)

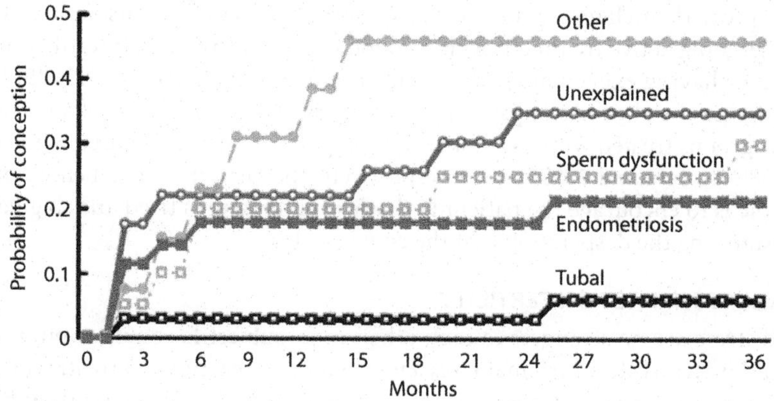

Figure 7.3 Cumulative conception rates without treatment analysed by diagnosis in couples who had completed IVF treatment at least 36 months previously. (From Cahill DJ et al. *Journal of Assisted Reproduction and Genetics*. 2005;22(11–12):401–5, with permission.)

either planning to have some treatment (IVF or IUI) or had already had some treatment (again IVF or IUI) but had not been successful. In their paper, for women aged less 35 years, 35–39 years and 40 years or more, 16%, 11% and 10% (respectively) delivered following natural conception (34).

7.3 ROLE OF ALTERNATIVE AND COMPLEMENTARY THERAPIES

Some couples use alternative and complementary therapies to supplement or treat their inability to conceive. Alternative and complementary therapies include 'homeopathic and herbal medicines, acupuncture and acupressure, energy healing, massage, specified diets, and psychosocial interventions' (36).

Population use of alternative and complementary therapies has increased worldwide to more than 50% across the developed world (37). Amongst infertile couples, one study found that almost 30% used some kind of alternative approach. Acupuncture was the most frequently used, followed by herbal remedies, meditation or body work, for example, massage (38). These options tend to be used by those in higher income brackets and those with some pre-existing belief in these therapies (38).

How effective are these alternative and complementary therapies? A detailed and good systematic review of benefit in infertility found 37 RCTs addressing acupuncture to relieve pain and to improve outcomes of IVF; the effect of supplements (selenium, zinc, ginseng and antioxidant vitamins, and for women only, phytoestrogens) on semen quality and function, and on clinical pregnancy rates; the effect on successful pregnancy rates following psychotherapy – and, unexpectedly, the benefits of weight loss (39). (One may be surprised that psychotherapy and weight loss are considered alternative and complementary therapies, but the evidence of their effect is useful.) Their recommendations centred on:

- Acupuncture (improvement in total motile sperm counts but pregnancy rates not studied; improved testicular blood flow following acupuncture – of uncertain value in fertility)
- Selenium supplementation with N-acetylcysteine (associated with increased sperm concentration, motility and morphologically normal sperm)
- Weight loss (with a 1000-kcal diet, weight loss with major positive biochemical differences, but hard outcomes such as ovulation or pregnancy rates were not reported)
- Psychotherapeutic intervention (some evidence for increased pregnancy rates following intervention) had three or more studies demonstrating beneficial effect (39)

These interventions should therefore be used, if appropriate to the couple's requirements. Patients who are desperate for anything to help them conceive and have a child will take anything and do anything to improve their chances. The role of psychotherapy is worth considering in the situation of couples undergoing active treatment and in couples with unexplained infertility, whether planning treatment or having expectant management. The review by Clark in 2013 (39) found several studies showing beneficial effects of psychotherapy in the hardest measure of all, pregnancy rates (40–42). In the groups treated with psychotherapy, pregnancy rates (compared with controls) were 54% vs 20%, 38% vs 14% and 28% vs 15%, respectively, the last set of data being nonsignificant). The doctor's role is to encourage the patient to disclose all measures tried and to give professional advice, understanding the desperation that the couple feels.

7.4 STRESS AND ITS IMPACT ON FERTILITY

Infertility and stress are no strangers to each other; most subfertile couples would agree that the quest for a child is stressful. A systematic review of women's reactions to treatment for infertility concluded that unsuccessful treatment was the principal factor in raising stress and anxiety, and that at the outset, the treatment group differed only slightly from controls (43). Nevertheless, the impact of stress on couples undergoing investigation and treatment is such that some fertility regulators insist that counselling and counsellors be available for couples having assisted conception treatment (e.g. the HFEA in the United Kingdom). But does stress play a *causative* role in infertility?

Lynch and colleagues in the United States examined salivary cortisol and alpha-amylase (both recognised markers of stress) in women prior to and during attempts at conception. They found that women with the highest levels of alpha-amylase had reduced probability of conception, with odds ratios for conception of 0.71 (95% CI 0.51–1.00) (i.e. a 29% reduction in the chance of conception) (44). Cortisol showed no differences. This is the first US study to publish this finding, and they replicated this in women in a British population (45). Neither study found a difference in cortisol levels, and subsequently, the reliability of alpha-amylase has been called into question (46). We tested the relationship between stress and infertility by measuring stress in women undergoing IVF treatment. Stress was measured in two ways – biomarkers which reflect physiological stress, and a validated questionnaire, the State-Trait Anxiety Inventory. The study collected serum and urine markers of stress (prolactin and cortisol) as well as State-Trait Anxiety Inventory data on women undergoing *in vitro* fertilisation treatment, both those stimulated with gonadotrophins and those with unstimulated natural cycles. They were compared with a control group of women undergoing gynaecological surgery unrelated to infertility (47). Stress biomarkers and anxiety levels were similar across and within all groups, indicating stress did not contribute greatly to infertility in the two affected groups. Women who managed to conceive had reduced state anxiety scores, but not significantly so.

Approaching the issue from another angle, the impact of measures to reduce negative emotions was tested in an RCT of brief, self-administered cognitive intervention compared with routine care, in 166 women starting routine IVF treatment (48). The cognitive intervention showed better psychological balance in the participants but had no impact on the chances of conception. In contrast, in 2000, later work by the same group found a positive effect on conception rates amongst those who had 10 sessions of group therapy (40).

In an extensive review of the effects of stress on infertility and infertility on stress, Rooney and Domar reviewed many papers and concluded that 'psychological interventions for women with infertility have the potential to decrease anxiety and depression and may well lead to significantly higher pregnancy rates' (49). Clark's systematic review, already referred to (39), found sufficient evidence to list psychotherapeutic approaches as a useful adjunct for couples having infertility treatment, with a possible beneficial effect on pregnancy rates.

Do ongoing traits of anxiety and stress or do environmental stress and negative life events negatively affect fecundity? One imagines that all these factors would play an adversarial role in conception and successful pregnancies, and it is a frequent plot device that a tremendous shock causes a woman to lose her baby. Given the delicate nature of the hypothalamic-pituitary- ovarian axis and the vulnerability of sperm production, one imagines that war, fear, hardship, trauma would affect fecundity – but surprisingly, no hard evidence has emerged to substantiate this belief.

So, how do we advise our patients? The stresses related to work and commuting to work will probably not be amenable to change. Psychotherapy is likely to reduce stress and may even help women get pregnant. Given that there is *some* evidence for that approach, it seems worth supporting that as an intervention.

7.5 LIFESTYLE CHOICES WHICH AFFECT FERTILITY

There are many choices we make in life which affect us in ways we don't expect. There are actions we take which affect our fertility to a greater or lesser extent, and we only become aware of the consequences when it is too late to do anything about it. These choices include deciding when in our lives we will have children; what our sexual behaviour is to be; whether we will smoke, drink alcohol, become and stay overweight; whether we take recreational/illegal drugs and the environments in which we live and work. The amount of published data in this area is not huge, but one US study prospectively surveyed 12,800 embarking on assisted conception. Of the women, 20% drank alcohol, 5% smoked, 64% drank caffeine, less than 1% used recreational drugs and 55% exercised during their IVF treatment.

Most of the morbidity I have seen in my career was caused by behaviour and choices people made when they were younger. It's not surprising then that I feel strongly about the importance of preventive measures in medicine. If men and women always used condoms as well as a contraceptive method, and if they decided to have children even 5 years earlier than they currently do, referrals to reproductive medicine services would decrease considerably, probably by half.

7.5.1 Timing of having children

For some individuals, when to have a child is not a choice they make, as they may get pregnant unexpectedly or in an unplanned way. For many people embarking on a relationship, having children along the way is part of the plan. Contraception has given humans incredible control over when to start a family, allowing the individual or couple time to gain qualifications and advance their career, travel, get decent accommodation and earn a salary which can cover child care costs. Many women are in their early 30s before they feel sufficiently secure to start a family.

Female fertility is at its peak in a woman's late teens and early 20s, but bearing children is not a priority for many women in their early 20s. Unfortunately, most women do not understand – or are unrealistic about – the biological norms of their own body. University undergraduates in the United States (average age 20.4 years [SD 2.3]) said that they wanted to have children in the future (positive responses – 88% female and 90% male). They were then asked at what age women's fertility declined. Thirty-six percent of females and 29% of males said it was 40–44 years, another 31% females and 52% males said it was 45–60 years (50). It is worrying that so many women – over 60% in this survey – and even more men, have such an erroneous view of their chances of having a child. When 2000 Canadian women aged 20–50 years were asked about their expectations for childbearing (51), the responses showed an awareness that they would probably become mothers later in life than what they considered to be ideal. They thought that 27 years was the ideal age for a woman to bear her first child, but they did not expect to have a child until they were 32. Men also need to be educated about the possibility of sperm deterioration over time. Men as well as women are putting off the opportunity to get pregnant early in life when physically and hormonally their bodies are best suited to conceiving (this simplistic presentation of that concept is discussed more fully in Section 9.2.3 in Chapter 9). Both men and women have a falsely optimistic view of how their fertility changes over time.

In response to this, what can we do as doctors? There is a move to educate the public about the realities of fertility, but the education needs to start in the early teenage years. Raising awareness in the media (52), particularly in magazines online and papers which attract a readership in the late teens and early 20s is also necessary. Some public health campaigns, warning about the consequences of waiting too long, have proved unpopular, for example, the Italian Fertility Day in 2016

(53), since they do not acknowledge the economic pressures on people in their twenties who cannot afford to pay child care costs and who lack job security. As with many issues to do with health, the economic and political environment has more effect on the nation's health than doctors alone can influence. The message is unpopular but important to broadcast.

7.5.2 The impact of unprotected sexual penetration

The impact of STIs on fertility is a matter of great concern to this author. There is a crisis in UK sexual health, where infection rates are rising inexorably. A national screening programme for sexually transmitted diseases, operating in the United Kingdom since 2003, shows rates of most STIs are continuing to rise. HIV and hepatitis B are included with sexually transmitted diseases but are not also counted when surveying incidence and prevalence of STIs. Between 2016 and 2017 in the United Kingdom, there were increases in diagnoses for gonorrhoea of 22%, from 36,577 to 44,676, and for syphilis (primary, secondary and early latent stages) a rise of 20%, from 5955 to 7137 (54). The numbers of all cause STIs had risen year on year for several years before 2016/17. A similar situation of increasing incidence exists in the United States, where half of all STIs occur in the 15–24 age group, and the rise (from 2016 to 2017) is 7% in cases of Chlamydia, 19% for gonorrhoea and 10.5% for syphilis (55).

Gonorrhoea, syphilis and the mycoplasmas, as well as hepatitis B and HIV, cause significant morbidity for those infected and, in some instances, their unborn babies, but these conditions generally have signs and symptoms, even if they are often ignored. Chlamydia is frequently silent when it is contracted, and problems only emerge later in the form of pelvic inflammatory disease and often in fallopian tube damage and problems with conception.

One strategy in the prevention of STIs is to encourage condom use. There have been numerous studies around the world examining sexual practice and the use of 'protection' during sexual activity. One common feature in these studies is the conflation of contraceptive precautions and barrier protection against STIs. Several studies report on both aspects as if they were one and the same thing. The data given below draw on any information provided which details barrier protection. Generally, they are not encouraging, and illustrate why the number of new STI cases is rising year on year.

One small survey (n = 57) held in-depth interviews with third-level students in Californian colleges (56). Thirty percent reported using no protective measures at all when having sex, 28% used condoms and 44% reported nonbarrier methods of contraception. More than one-third of those who were sexually active had had unprotected sex in the previous 3 months. An earlier report from a midwestern university found that, during their first sexual encounter, many students (women, 63.2%; men, 57.4%) did not use any contraceptive. Of those who did, most used condoms (82.9% and 52.5%, respectively) (57).

A group of 2400 young (aged 14–24 years) people in a Portuguese city were surveyed about sexual and contraceptive behaviour (58). By the time they were surveyed, 58% of people in the sample had already had sex. The mean age at the first sexual intercourse was 16.4 ± 1.8 years. The condom was the contraceptive method of choice for planned and unplanned first and subsequent encounters (58).

In Ireland, a self-completed questionnaire data was collected from 4494 young people, 15–18 years old, enquiring about different aspects of their behaviour, including sexual debut and contraceptive use (59). At the time of the survey, 20%–25% had already had their first sexual encounter – a surprisingly low number. Condom use was reported by 80% of sexually initiated students at the last intercourse, but inquiry was not made into condom use during the entire period of sexual activity (59).

In the United Kingdom, data from all people presenting to emergency contraception clinics suggest that clients with a mean age 26 years (range 15–49 years) used condoms in 42% of cases only, and nothing at all in 47% (60).

Preventing STIs is a public health issue and requires the combined efforts of the individual citizen, public education about safe sex and screening clinics. With this in mind, and alarmed by the rising statistics, the United Kingdom launched a public education campaign about STIs in 2017, aimed at teenagers and young adults, and encouraging condom use, amongst other things. They have produced two excellent videos which can be accessed online from reference (61).

Similar campaigns have been implemented in other countries. The question is: How effective are these campaigns, and what really works to influence young people to adopt safer practices? There are key points in any education/information intervention as listed here.

Every additional sexual partner increases your risk of getting an STI. Know your sexual partners and limit their number – their sexual history is as important as your own.

Get checked for STIs when you enter a new relationship, and at regular intervals if you or your partner has sex with other people.

Plan and prepare if you are likely to engage in sexual activity with alcohol or drugs – have a condom with you and use it. If you have unprotected sex in this way, get tested.

Use only latex condoms every time you have vaginal, oral or anal sex; this decreases the chances of infection. Frequent use of some spermicides on condoms can increase the risk of HIV.

Be sure you know how to put on a condom correctly; watch videos to check your technique.

Be aware of the risks of different sex practices. Be aware that anal sex poses an especially high risk because tissues in the rectum tear easily, which makes STIs more easily transmitted. But even oral sex can transmit STIs.

Get immunised; vaccinations will help prevent hepatitis B and some types of HPV.

A 2016 systematic review of the effectiveness of education interventions for adolescents in the United States examined only programmes which included STI checks before and afterwards. Reduction in STIs was the outcome of interest. Abstinence-only programmes were ineffective on this measure. Educational interventions promoting safe sexual practice, teaching communication and negotiation skills and showing how to use condoms did have a positive effect (62). Similar approaches, tailored to the specific environment of working in low- and middle-income countries, were reported in a 2016 review (63). The authors considered their results of significance to those who wanted to develop strategic plans in this area.

7.5.3 Body mass index outside the normal range

The normal range for body mass index is 19–25 kg/m^2. Fertility is adversely affected by increasing BMI, as outlined in Chapter 2 (Section 2.1.2). An extensive WHO supported study has examined body mass index data from 29 developed and 140 developing countries (64). Global BMI levels have increased between 1980 and 2008 by 0.4 kg/m^2/decade, with men having a greater increase in BMI by 0.9–1.1 kg/m^2/decade. It is possible to look at the relative change in BMI for any country over the 1980–2008 interval. The increase in Western Europe was less than 0.4 in females, and 0.6 in men (kg/m^2/decade), compared with the increase in North America, where BMI increased by about 1.1 in men and 1.2 in females. Nauru (Micronesia) had the highest average BMI in 1980, and still held this position in 2008 (for both men and women, 34 and 35, respectively); Bangladesh and Vietnam exchanged places for the lowest BMI in 1980 and 2008 (men and women, 19.9 and 20.5, respectively) (64).

It is clear from these data that the world is gaining weight, and as it does, the impact of increase on fertility will become more and more evident. Individual publications and national guidelines all support the point that fertility is diminished as weight (and BMI) increases (65,66). The significance of the effect is such that in the United Kingdom, BMI above 30 is one of the limiting factors (treatment will not be offered) on fertility treatment funded by the National Health Service. Similar criteria (for weight) do not appear to be applied in Europe (67), while in the United States, the position is similar (no regulation), though clinicians are beginning to say they should be – recognising the challenges in success rates and in pregnancy problems if conception occurs (68).

If a couple wants advice in this regard, even if you are not working in a clinic with weight restrictions, the woman particularly needs to lose weight. All the traditional methods can be used by joining a weight reduction programme, but if that is having no obvious effect after several months, appetite suppressants could be tried (but are generally not used as they have very severe side effects – suicidal ideation and cerebrovascular accidents) and a safer option is orlistat, which prevents the absorption of fat, and when used compliantly is associated with weight loss (though not enormous – up to 35%–73% for orlistat) (69).

Two recent studies in the Netherlands and in Sweden may serve to turn all the perceived wisdom above on its head. The Dutch study randomised women (with a BMI of 29 or more) into a lifestyle-intervention programme (1200 cal intake/day and 10,000 steps/day) and a control group (70). The intervention group lost an average of 4.4 kg. The live birth rate in the intervention group (27%) was less than that in the control group (35%) (RR 0.77; 95% CI, 0.60–0.99). The Swedish study was similar, though the intervention was more severe and focussed more on weight loss than lifestyle (71). They studied woman with BMIs between 30 and 35. The intervention group had a daily caloric intake of 880 kcal. In this study, the live birth rate in the intervention group (29%) was no different to that in the control group (27%). The intervention group lost an average of 9.4 kg. Commenting elsewhere on these studies, Norman and Mol review the data and conclude that the effect of these results should be that clinicians should consider moving to fertility treatment earlier than originally recommended for patients who are overweight or obese, and not wait for people to lose weight, as it appears to have little impact of the outcome of live birth (72).

7.5.4 Legal and illegal substances which affect fertility

In Chapter 4, there is a section which discusses the effects of smoking, alcohol and illicit drugs on fertility in men.

The effect of *cigarette smoking* on female fertility was investigated intensively 10–20 years ago. The evidence was so compelling that there haven't been studies in this area for several years. There is both laboratory (73,74) and population-based (75) evidence that cigarette smoking reduces fertility. Despite this, at least 5% of the infertile population smokes and 20% of infertile women drink alcohol (76). A recent Australian study used weekly diaries to look at the impact of lifestyle behaviours on IVF outcome (77). Confirming previous findings, male smoking led to a poor outcome (as an increased risk of pregnancy loss) and female smoking (in years smoked) had a direct effect on basal FSH levels as a marker of ovarian reserve (77). Basal FSH levels are directly related to numbers of mature follicles and oocyte yield (78).

The evidence for an effect of *alcohol* on getting pregnant is less convincing (79). In their data, Fims et al. suggested that moderate intake in men and in women was associated with higher fertilisation rates in the partners of the men and no effect in the women. Data collected from the Environment and Reproductive Health Study, a prospective cohort study, focussed on women in the study having assisted conception (79). In that, the investigators found that there was no difference in live births between women who had no alcohol and those consuming up to 2.4 g/day (i.e. a lot – depending on the country you live in, 20–30 units/day).

Questions have been raised about the effect of *caffeine* on conception, but, as with alcohol, there is no convincing evidence of this (79,80). *Cannabinoids* in marijuana do alter hypothalamic-pituitary-ovarian axis function in human and animal studies by suppressing GnRH activity and thus bringing about an overall suppression of ovarian and ovulatory function (81). Most of the evidence of the effects of cannabinoids comes from animal studies – there are no RCTs I could find on cannabinoid effect on fertility. Nonetheless, several effects on reproduction have been observed. Endogenous and exogenous cannabinoids have been demonstrated to alter tubal motility, which leads to the possibility that they may have a role to play in tubal pregnancy and, further, they inhibit embryo development at the two-cell stage (82). One of the two most studied endocannabinoids, arachidonyol-ethanolamine (AEA) is suspected to regulate the implantation 'window' in the uterus and, further, they inhibit embryo development at the two-cell stage (82). In men, cannabinoids affect spermatogenesis, though the mechanism is unclear. It is thought to be modulated by reducing testosterone levels (82). It would appear that *opiates* have a similar effect on GnRH activity

(83). The finding of reduced sperm quality was replicated recently, but Murphy and colleagues also found worrying evidence of DNA methylation damage in the sperm of users of cannabis (with blood levels of tetrahydrocannabinol ranging from 50 to 739 ng/mL) (84).

Advice: couples should clearly stop smoking – that has negative effects in men and women. For alcohol, they should not exceed their daily limits to avoid fetal alcohol syndrome. No convincing effect on pregnancy rates exist, but many couples are well into the first trimester before they realise the woman is pregnant. There seems to be no reason to reduce caffeine prior to pregnancy. Marijuana use is likely to have some effects on the likelihood of conception – reduced sperm concentration and one report suggesting alteration of DNA function. Stopping it may help, particularly is there is little else wrong with a couple.

7.5.5 Mobile phones and (Wi-Fi) computer use

In Chapter 4 (Section 4.5), whether mobile phones and computers (specifically laptops with Wi-Fi) have an effect of fertility was addressed. Animal studies suggested mobile phone usage decreases sperm count and motility and increases oxidative stress (85). Human studies estimated sperm motility is reduced by approximately 8% in exposed compared to nonexposed groups (86). A laboratory study on sperm aliquots at constant temperature allocated one-half to Wi-Fi exposure from a laptop, the other half not. In the exposed group, a decrease in progressive sperm motility and an increase in sperm DNA fragmentation was observed (87).

Much of the evidence about mobile phones and (Wi-Fi) computer use relates to the effect on men and sperm. Speculation about oxidative stress and mitochondrial damage in women in oocytes appears to have little foundation as yet, as all these studies are in rat and mouse model experiments (94). It is possible that further investigation will reverse this.

Should any of this prompt a change in behaviour? I think that if there is an obvious cause for a couple's fertility, the impact of wireless signals from mobile phones and laptops will have only a marginal effect. If they have unexplained infertility, a change in behaviour might be worthwhile, recognising it will need to be for 3–4 months as a minimum.

7.5.6 Prescribed and over-the-counter medications

There is a lot of discussion about the possible negative impact of nonsteroidal anti-inflammatory analgesics because of concern about prostaglandin inhibitors. There are considerable animal data to support this view (95). In humans, one prospective cohort study of over 2000 women found a dose–response relation between naproxen use and the ability to conceive; the difference using <1500 and ≥1500 mg of naproxen was 0.85 (95% CI: 0.68–1.07) (reduced but not statistically significant) (96). In a smaller study of 259 women, the women were followed up for two menstrual cycles, and data on nonsteroidal anti-inflammatory drugs (NSAIDs) (ibuprofen, acetaminophen [paracetamol], aspirin and naproxen) were collected. The conclusion was that NSAIDs were unlikely to harm reproductive function (97). Nonsteroidal anti-inflammatory analgesics probably have no clinical impact on the likelihood of getting pregnant.

In Chapter 5, antipsychotics and H2 blockers were mentioned, which are recognised to have a detrimental effect on fertility, through their influence on prolactin levels or weight gain (92). If the female partner has to take these medications, she should continue to do so. These drugs may affect natural conception, but once exogenous gonadotrophins are introduced, any effects of prolactin will be overruled by the gonadotrophin therapy. A 2019 review of selective serotonin reuptake inhibitors (SSRIs) examined 16 original papers on SSRI impact on fertility (88). Of seven studies in women, six papers found no significant association between SSRIs and treatment outcomes; in men, a similar number of papers were found to have an adverse effect on semen parameters. In men, sperm numbers are reduced, and DNA integrity damage increased. In addition, problems of reduced libido, erectile dysfunction and delayed ejaculation are noted (88).

Bear in mind that even homeopathic medicine can have a detrimental effect on ovarian function. The unknown content of the medication will often make it difficult to define the mechanism of action, but I have seen such medications cause grossly abnormal oestradiol and gonadotrophin levels in one woman (90). In any pharmacy, there will be a range of products available which are promoted to boost fertility, sperm quality and female endocrinology. Most of these are not supported by convincing scientific evidence. A recent conference contribution presented data of the relationship between sperm quality and supplements (91). This study was underpowered and so only interferences can be drawn, not deductions. Men were surveyed as they entered the clinic's care and any history of dietary supplements was noted. Of the men surveyed, 54 had sperm samples analysed and there were no differences noted between men taking supplements and controls (not taking supplements). They also noted, interestingly, that men from wealthier backgrounds or less socially deprived areas were more likely to take supplements. When supplements contain antioxidants (e.g. vitamin C, zinc), they may have some benefit on sperm quality, but how much gets to the seminiferous tubules in the testis is difficult to say.

I think that any medication a man or woman is taking could affect their fertility. If fertility is in question, you as a clinician should check the data on the medication to check if there is any contraindication and advise accordingly. Nonsteroidal anti-inflammatory drugs do not appear to affect ovulation, despite the theoretical risks. Complementary medicines are in my view more unpredictable and therefore more risky from their potential effects on reproductive function. I would avoid them in the month before and during any treatment involving gonadotrophin therapy.

7.6 CASE STUDY

Janet and Peter have been married for 5 years, and stopped using contraception a year after they married, when Janet was 29. Two years ago, they asked to be referred by their GP to see a gynaecologist, as Janet had not become pregnant. Her sister had blocked tubes, and they were concerned that any delay in diagnosis might make things more difficult for them. After 6 months of blood and sperm tests, an x-ray (a hysterosalpingogram) of Janet's uterus and a laparoscopy, they were told that there was no obvious cause for their failure to get pregnant. The gynaecologist was positive and encouraging and said that, as this was the case, they had every chance of getting pregnant by themselves. The advice was they should relax, have regular intercourse and wait for it to happen.

The couple went on a holiday to Spain, and they went away for lots of weekends, but Janet still didn't get pregnant. After a year, Janet suggested going back to see the gynaecologist. By then, they were becoming desperate. The gynaecologist went over things again, explained how everything was in good order and that no simple measure or treatment was going to help improve their chances. The only alternative to waiting for pregnancy to occur naturally was to intervene with IVF. After giving this some thought and working out that they would have one health service–funded cycle and could afford one further cycle of treatment, they asked to be referred to the nearest IVF clinic, 40 miles away. When they went there, it seemed as if a lot of the tests were repeated. Within a few weeks, Janet had an egg collection operation, and then an embryo transfer. Unfortunately, her period came 2 weeks later. They had a visit to the clinic to discuss the cycle and plan the next. She is going to have this soon, with no change to the treatment plan.

7.6.1 Commentary

Infertility does not run in families, although sperm disorders may be passed on and endometriosis and PCOS are more commonly found in other family members than in the general population. Tubal damage, however, does not, as it tends to be related directly to the sexual behaviour of the individuals. The diagnosis in this case was essentially 'unexplained infertility', although this strictly requires 3 years of infertility. The role of factors such as stress is poorly understood and difficult to define. Whenever formal studies of stress levels and stress chemicals are undertaken, no major differences are found in the infertile population, though psychological interventions may be of help. Yet there are many anecdotes of people who give up trying, adopt and then find themselves pregnant. At the end of 2 years, a period sometimes called expectant waiting, 95% of couples will have conceived and, in the next year, there is still a likelihood of conception, though increasingly small. However, after 3 years, the chances of natural conception are very low and, realistically, the best chances then are from treatments such as IVF. Many IVF clinics will want to repeat some or all the tests that are relevant. This is partly to have accurate results, partly to ensure nothing dramatic has changed since initial testing and partly because some of the tests have important predictive value for IVF outcome. In IVF for couples with unexplained infertility, the expected success rate is about 30% per cycle, compared with the 1%–2% rate per cycle that is otherwise seen after 3 years of trying to conceive naturally. With three cycles of IVF, an overall success rate of over 60% should be expected if the woman is less than 40 years old.

7.7 SUMMARY

Unexplained infertility is a frustrating diagnosis for couples. For clinicians, there is a fairly rigourous set of criteria to be met to justify the diagnosis, which means that all the appropriate investigations have been done to rule out tubal, ovulatory and sperm disorders. Endometriosis is a separate diagnosis which causes subfertility even in minor disease and which can be treated by surgical ablation. Alternative therapies offer little to improve fertility, and when stress is measured chemically or by questionnaire, there is little evidence for its role in being causative as a cause of infertility.

There is much patients can do themselves to improve fertility. Clearly, women cannot get younger, but other patterns of behaviour referred to in this chapter and elsewhere (Section 4.5 in Chapter 4) can be reversed (though not easily) and are likely to promote fertility if those changes can be achieved. These would be

- Stopping smoking for both men and women
- Minimising alcohol to recommended levels or lower
- A loss of weight for women (though data in this chapter in some way negate that advice and recommend more immediate treatment)
- For completely unexplained infertility, minimising exposure to mobile phones and laptops
- Psychotherapy is of value, and worth undertaking, given the evidence base for it
- Recreational drugs (particularly cannabis) are detrimental to fertility and it makes sense to stop them if used by the man or the woman

REFERENCES

1. Hull MG, Cahill DJ. Female infertility. *Endocrinology and Metabolism Clinics of North America*. 1998;27(4):851–76.
2. Hull MG, Glazener CM, Kelly NJ, Conway DI, Foster PA, Hinton RA et al. Population study of causes, treatment, and outcome of infertility. *British Medical Journal (Clinical Research ed)*. 1985;291(6510):1693–7.
3. Oei SG, Helmerhorst FM, Bloemenkamp KW, Hollants FA, Meerpoel DE, Keirse MJ. Effectiveness of the postcoital test: Randomised controlled trial. *BMJ (Clinical Research Ed)*. 1998;317(7157):502–5.
4. Kamath MS, Bhattacharya S. Demographics of infertility and management of unexplained infertility. *Best Practice & Research Clinical Obstetrics & Gynaecology*. 2012;26(6):729–38.
5. Revised American Society for Reproductive Medicine classification of endometriosis: 1996. *Fertility and Sterility*. 1997;67(5):817–21.
6. Mahmood TA, Arumugam K, Templeton AA. Oocyte and follicular fluid characteristics in women with mild endometriosis. *British Journal of Obstetrics and Gynaecology*. 1991;98(6):573–7.
7. Wardle PG, Mitchell JD, McLaughlin EA, Ray BD, McDermott A, Hull MG. Endometriosis and ovulatory disorder: Reduced fertilisation *in vitro* compared with tubal and unexplained infertility. *Lancet (London, England)*. 1985;2(8449):236–9.

8. Mills MS, Eddowes HA, Cahill DJ, Fahy UM, Abuzeid MI, McDermott A et al. A prospective controlled study of *in-vitro* fertilization, gamete intra-fallopian transfer and intrauterine insemination combined with superovulation. *Human Reproduction (Oxford, England).* 1992;7(4):490–4.
9. Cahill DJ, Wardle PG, Maile LA, Harlow CR, Hull MG. Pituitary-ovarian dysfunction as a cause for endometriosis-associated and unexplained infertility. *Human Reproduction (Oxford, England).* 1995;10(12):3142–46.
10. Akande VA, Hunt LP, Cahill DJ, Jenkins JM. Differences in time to natural conception between women with unexplained infertility and infertile women with minor endometriosis. *Human Reproduction (Oxford, England).* 2004;19(1):96–103.
11. D'Hooghe TM, Debrock S, Hill JA, Meuleman C. Endometriosis and subfertility: Is the relationship resolved?. *Seminars in Reproductive Medicine.* 2003;21(2):243–54.
12. Cahill DJ, Wardle PG, Harlow CR, Hull MG. Effect of progestogen therapy on follicular development, related hormone concentrations and fertilization *in vitro* in unstimulated cycles and unexplained and endometriosis-associated infertility. *Human Reproduction (Oxford, England).* 1996;11(3):647–50.
13. Duffy JM, Arambage K, Correa FJ, Olive D, Farquhar C, Garry R et al. Laparoscopic surgery for endometriosis. *Cochrane Database of Systematic Reviews.* 2014:4:Cd011031.
14. Cahill DJ, Wardle PG, Maile LA, Harlow CR, Hull MG. Ovarian dysfunction in endometriosis-associated and unexplained infertility. *Journal of Assisted Reproduction and Genetics.* 1997;14(10):554–7.
15. Omland AK, Abyholm T, Fedorcsak P, Ertzeid G, Oldereid NB, Bjercke S et al. Pregnancy outcome after IVF and ICSI in unexplained, endometriosis-associated and tubal factor infertility. *Human Reproduction (Oxford, England).* 2005;20(3):722–7.
16. Bancroft K, Vaughan Williams CA, Elstein M. Pituitary-ovarian function in women with minimal or mild endometriosis and otherwise unexplained infertility. *Clinical Endocrinology.* 1992;36(2):177–81.
17. Tummon IS, Maclin VM, Radwanska E, Binor Z, Dmowski WP. Occult ovulatory dysfunction in women with minimal endometriosis or unexplained infertility. *Fertility and Sterility.* 1988;50(5):716–20.
18. Bérubé S, Marcoux S, Langevin M, Maheux R. Fecundity of infertile women with minimal or mild endometriosis and women with unexplained infertility. *Fertility and Sterility.* 1998;69(6):1034–41.
19. Jansen RP. Minimal endometriosis and reduced fecundability: Prospective evidence from an artificial insemination by donor program. *Fertility and Sterility.* 1986;46(1):141–3.
20. Gonzalez-Comadran M, Schwarze JE, Zegers-Hochschild F, Souza MD, Carreras R, Checa MA. The impact of endometriosis on the outcome of Assisted Reproductive Technology. *Reproductive Biology and Endocrinology: RB&E.* 2017;15(1):8.
21. Fleming CF, Hughes AO, Williams JA, Hull MG. Effect of cause of infertility on success during IVF. *Human Reproduction (Oxford, England).* 1994;9(Suppl. 1):117.
22. Hull MG, Eddowes HA, Fahy U, Abuzeid MI, Mills MS, Cahill DJ et al. Expectations of assisted conception for infertility. *BMJ (Clinical Research Ed).* 1992;304(6840):1465–9.
23. Hull MG, Williams JA, Ray B, McLaughlin EA, Akande VA, Ford WC. The contribution of subtle oocyte or sperm dysfunction affecting fertilization in endometriosis-associated or unexplained infertility: a controlled comparison with tubal infertility and use of donor spermatozoa. *Human Reproduction (Oxford, England).* 1998;13(7):1825–30.
24. Marcoux S, Maheux R, Berube S. Laparoscopic surgery in infertile women with minimal or mild endometriosis. Canadian Collaborative Group on Endometriosis. *New England Journal of Medicine.* 1997;337(4):217–22.
25. Parazzini F. Ablation of lesions or no treatment in minimal-mild endometriosis in infertile women: A randomized trial. Gruppo Italiano per lo Studio dell'Endometriosi. *Human Reproduction (Oxford, England).* 1999;14(5):1332–4.

26. Tanrikut E, Karaer A, Celik O, Celik E, Otlu B, Yilmaz E et al. Role of endometrial concentrations of heavy metals (cadmium, lead, mercury and arsenic) in the aetiology of unexplained infertility. *European Journal of Obstetrics, Gynecology, and Reproductive Biology.* 2014;179:187–90.
27. Butts SF, Seifer DB, Koelper N, Senapati S, Sammel MD, Hoofnagle AN et al. Vitamin D deficiency is associated with poor ovarian stimulation outcome in PCOS but not unexplained infertility. *Journal of Clinical Endocrinology and Metabolism.* 2019;104(2):369–78.
28. Sadeghi MR. Unexplained infertility, the controversial matter in management of infertile couples. *Journal of Reproduction & Infertility.* 2015;16(1):1–2.
29. Yucel B, Demirel E, Kelekci S, Shawki O. Hysteroscopic evaluation of tubal peristaltic dysfunction in unexplained infertility. *Journal of Obstetrics and Gynaecology: Journal of the Institute of Obstetrics and Gynaecology.* 2018;38(4):511–5.
30. Taylor PJ. When is enough enough? *Fertility and Sterility.* 1990;54(5):772–4.
31. National Collaborating Centre for Women's and Children's Health. National Institute for Health and Clinical Excellence: Guidance. *Fertility: Assessment and Treatment for People with Fertility Problems. London: Royal College of Obstetricians & Gynaecologists.* National Collaborating Centre for Women's and Children's Health. 2013.
32. De Cicco S, Tagliaferri V, Selvaggi L, Romualdi D, Di Florio C, Immediata V et al. Expectant management may reduce overtreatment in women affected by unexplained infertility confirmed by diagnostic laparoscopy. *Archives of Gynecology and Obstetrics.* 2017;295(2):427–33.
33. Hughes E, Brown J, Collins JJ, Vanderkerchove P. Clomiphene citrate for unexplained subfertility in women. *Cochrane Database of Systematic Reviews.* 2010:1:Cd000057.
34. Malchau SS, Henningsen AA, Loft A, Rasmussen S, Forman J, Nyboe Andersen A et al. The long-term prognosis for live birth in couples initiating fertility treatments. *Human Reproduction (Oxford, England).* 2017;32(7):1439–49.
35. Cahill DJ, Meadowcroft J, Akande VA, Corrigan E. Likelihood of natural conception following treatment by IVF. *Journal of Assisted Reproduction and Genetics.* 2005;22(11–12):401–5.
36. Zollman C, Vickers A. What is complementary medicine? *BMJ (Clinical Research Ed).* 1999;319(7211):693–6.
37. Rayner JA, Willis K, Burgess R. Women's use of complementary and alternative medicine for fertility enhancement: A review of the literature. *Journal of Alternative and Complementary Medicine (New York, NY).* 2011;17(8):685–90.
38. Smith JF, Eisenberg ML, Millstein SG, Nachtigall RD, Shindel AW, Wing H et al. The use of complementary and alternative fertility treatment in couples seeking fertility care: Data from a prospective cohort in the United States. *Fertility and Sterility.* 2010;93(7):2169–74.
39. Clark NA, Will M, Moravek MB, Fisseha S. A systematic review of the evidence for complementary and alternative medicine in infertility. *International Journal of Gynaecology and Obstetrics: Official Organ of the International Federation of Gynaecology and Obstetrics.* 2013;122(3):202–6.
40. Domar AD, Clapp D, Slawsby EA, Dusek J, Kessel B, Freizinger M. Impact of group psychological interventions on pregnancy rates in infertile women. *Fertility and Sterility.* 2000;73(4):805–11.
41. Hosaka T, Matsubayashi H, Sugiyama Y, Izumi S, Makino T. Effect of psychiatric group intervention on natural-killer cell activity and pregnancy rate. *General Hospital Psychiatry.* 2002;24(5):353–6.
42. Chan CH, Ng EH, Chan CL, Ho, Chan TH. Effectiveness of psychosocial group intervention for reducing anxiety in women undergoing *in vitro* fertilization: A randomized controlled study. *Fertility and Sterility.* 2006;85(2):339–46.
43. Verhaak CM, Smeenk JM, Evers AW, Kremer JA, Kraaimaat FW, Braat DD. Women's emotional adjustment to IVF: A systematic review of 25 years of research. *Human Reproduction Update.* 2007;13(1):27–36.

44. Lynch CD, Sundaram R, Maisog JM, Sweeney AM, Buck Louis GM. Preconception stress increases the risk of infertility: Results from a couple-based prospective cohort study – The LIFE study. *Human Reproduction (Oxford, England)*. 2014;29(5):1067–75.
45. Louis GM, Lum KJ, Sundaram R, Chen Z, Kim S, Lynch CD et al. Stress reduces conception probabilities across the fertile window: Evidence in support of relaxation. *Fertility and Sterility*. 2011;95(7):2184–9.
46. Stegmann BJ. Other nonstress influences can alter salivary α-amylase activity. *Fertility and Sterility*. 2011;95(7):2190–1.
47. Harlow CR, Fahy UM, Talbot WM, Wardle PG, Hull MG. Stress and stress-related hormones during *in-vitro* fertilization treatment. *Human Reproduction (Oxford, England)*. 1996;11(2):274–9.
48. Domar AD, Gross J, Rooney K, Boivin J. Exploratory randomized trial on the effect of a brief psychological intervention on emotions, quality of life, discontinuation, and pregnancy rates in *in vitro* fertilization patients. *Fertility and Sterility*. 2015;104(2):440–51.e7.
49. Rooney KL, Domar AD. The relationship between stress and infertility. *Dialogues in Clinical Neuroscience*. 2018;20(1):41–7.
50. Peterson BD, Pirritano M, Tucker L, Lampic C. Fertility awareness and parenting attitudes among American male and female undergraduate university students. *Human Reproduction (Oxford, England)*. 2012;27(5):1375–82.
51. Daniluk JC, Koert E. Childless Canadian men's and women's childbearing intentions, attitudes towards and willingness to use assisted human reproduction. *Human Reproduction (Oxford, England)*. 2012;27(8):2405–12.
52. Lindsay J. At what age are women most fertile? Metro. 2018 Wednesday 14 Nov 2018.
53. Scherer S. Italian minister under renewed fire for Fertility Day campaign London, England: Reuters; 2016 [cited 2016]. Available from: https://www.reuters.com/article/us-italy-politics-fertility/italian-minister-under-renewed-fire-for-fertility-day-campaign-idUSKCN11S1KY.
54. Public Health England. *Sexually Transmitted Infections and Screening for Chlamydia in England, 2017. Health Protection Report*. 2018;12(20):28.
55. Braxton J, Davis D, Emerson B. Sexually Transmitted Disease Surveillance 2017. Atlanta: U.S. Department of Health and Human Services; 2018. Available from: https://www.cdc.gov/std/stats17/default.htm.
56. Cabral MA, Schroeder R, Armstrong EM, El Ayadi AM, Gurel AL, Chang J et al. Pregnancy intentions, contraceptive knowledge and educational aspirations among community college students. *Perspectives on Sexual and Reproductive Health*. 2018;50(4):181–8.
57. Darling CA, Davidson JK, Sr., Passarello LC. The mystique of first intercourse among college youth: The role of partners, contraceptive practices, and psychological reactions. *Journal of Youth and Adolescence*. 1992;21(1):97–117.
58. Miranda PSF, Aquino JMG, Monteiro R, Dixe M, Luz A, Moleiro P. Sexual behaviors: Study in the youth. *Einstein (Sao Paulo, Brazil)*. 2018;16(3):eAO4265.
59. Young H, Burke L, Nic Gabhainn S. Sexual intercourse, age of initiation and contraception among adolescents in Ireland: Findings from the Health Behaviour in School-aged Children (HBSC) Ireland study. *BMC Public Health*. 2018;18(1):362.
60. Cameron ST, Glasier A, Johnstone A, Rae L. Ongoing contraception after use of emergency contraception from a specialist contraceptive service. *Contraception*. 2011;84(4):368–71.
61. Thompson S. An STI is diagnosed in a young person every 4 minutes in England London: An STI is diagnosed in a young person every 4 minutes in England; 2018. Available from: https://www.gov.uk/government/news/an-sti-is-diagnosed-in-a-young-person-every-4-minutes-in-england.
62. Petrova D, Garcia-Retamero R. Effective evidence-based programs for preventing sexually-transmitted infections: A meta-analysis. *Current HIV Research*. 2015;13(5):432–8.

63. Kalamar AM, Bayer AM, Hindin MJ. Interventions to prevent sexually transmitted infections, including HIV, among young people in low- and middle-income countries: A systematic review of the published and gray literature. *Journal of Adolescent Health: Official Publication of the Society for Adolescent Medicine*. 2016;59(3 Suppl):S22–31.
64. Finucane MM, Stevens GA, Cowan MJ, Danaei G, Lin JK, Paciorek CJ et al. National, regional, and global trends in body-mass index since 1980: Systematic analysis of health examination surveys and epidemiological studies with 960 country-years and 9.1 million participants. *Lancet (London, England)*. 2011;377(9765):557–67.
65. National Collaborating Centre for Women's and Children's Health. National Institute for Health and Clinical Excellence: Guidance. Fertility: Assessment and Treatment for People with Fertility Problems. London: Royal College of Obstetricians & Gynaecologists National Collaborating Centre for Women's and Children's Health; 2013.
66. Wang JX, Davies M, Norman RJ. Body mass and probability of pregnancy during assisted reproduction treatment: Retrospective study. *BMJ (Clinical Research Ed)*. 2000;321(7272):1320–1.
67. Berg Brigham K, Cadier B, Chevreul K. The diversity of regulation and public financing of IVF in Europe and its impact on utilization. *Human Reproduction*. 2013;28(3):666–75.
68. Vahratian A, Smith YR. Should access to fertility-related services be conditional on body mass index? *Human Reproduction (Oxford, England)*. 2009;24(7):1532–7.
69. Yanovski SZ, Yanovski JA. Long-term drug treatment for obesity: A systematic and clinical review. *Jama*. 2014;311(1):74–86.
70. Mutsaerts MAQ, van Oers AM, Groen H, Burggraaff JM, Kuchenbecker WKH, Perquin DAM et al. Randomized trial of a lifestyle program in obese infertile women. *New England Journal of Medicine*. 2016;374(20):1942–53.
71. Einarsson S, Bergh C, Thurin-Kjellberg A, Mikkelsen-Englund A-L, Stenlöf K, Wistrand A et al. Weight reduction intervention for obese infertile women prior to IVF: A randomized controlled trial. *Human Reproduction*. 2017;32(8):1621–30.
72. Norman RJ, Mol BWJ. Successful weight loss interventions before *in vitro* fertilization: Fat chance?. *Fertility and Sterility*. 2018;110(4):581–6.
73. Rosevear SK, Holt DW, Lee TD, Ford WC, Wardle PG, Hull MG. Smoking and decreased fertilisation rates *in vitro*. *Lancet (London, England)*. 1992;340(8829):1195–6.
74. Zenzes MT, Reed TE, Casper RF. Effects of cigarette smoking and age on the maturation of human oocytes. *Human Reproduction (Oxford, England)*. 1997;12(8):1736–41.
75. Hull MG, North K, Taylor H, Farrow A, Ford WC. Delayed conception and active and passive smoking. The Avon Longitudinal Study of Pregnancy and Childhood Study Team. *Fertility and Sterility*. 2000;74(4):725–33.
76. Domar AD, Rooney KL, Milstein M, Conboy L. Lifestyle habits of 12,800 IVF patients: Prevalence of negative lifestyle behaviors, and impact of region and insurance coverage. *Human Fertility (Cambridge, England)*. 2015;18(4):253–7.
77. Firns S, Cruzat VF, Keane KN, Joesbury KA, Lee AH, Newsholme P et al. The effect of cigarette smoking, alcohol consumption and fruit and vegetable consumption on IVF outcomes: A review and presentation of original data. *Reproductive Biology and Endocrinology: RB&E*. 2015;13:134.
78. Cahill DJ, Prosser CJ, Wardle PG, Ford WC, Hull MG. Relative influence of serum follicle stimulating hormone, age and other factors on ovarian response to gonadotrophin stimulation. *British Journal of Obstetrics and Gynaecology*. 1994;101(11):999–1002.
79. Abadia L, Chiu YH, Williams PL, Toth TL, Souter I, Hauser R et al. The association between pre-treatment maternal alcohol and caffeine intake and outcomes of assisted reproduction in a prospectively followed cohort. *Human Reproduction (Oxford, England)*. 2017;32(9):1846–54.
80. Minguez-Alarcon L, Chavarro JE, Gaskins AJ. Caffeine, alcohol, smoking, and reproductive outcomes among couples undergoing assisted reproductive technology treatments. *Fertility and Sterility*. 2018;110(4):587–92.

81. Brents LK. Marijuana, the endocannabinoid system and the female reproductive system. *Yale Journal of Biology and Medicine*. 2016;27(89):175–91.
82. Taylor AH, Ang C, Bell SC, Konje JC. The role of the endocannabinoid system in gametogenesis, implantation and early pregnancy. *Human Reproduction Update*. 2007;13(5):501–13.
83. Böttcher B, Seeber B, Leyendecker G, Wildt L. Impact of the opioid system on the reproductive axis. *Fertility and Sterility*. 2017;108(2):207–13.
84. Murphy SK, Itchon-Ramos N, Visco Z, Huang Z, Grenier C, Schrott R et al. Cannabinoid exposure and altered DNA methylation in rat and human sperm. *Epigenetics*. 2018.
85. La Vignera S, Condorelli RA, Vicari E, D'Agata R, Calogero AE. Effects of the exposure to mobile phones on male reproduction: A review of the literature. *Journal of Andrology*. 2012;33(3):350–6.
86. Adams JA, Galloway TS, Mondal D, Esteves SC, Mathews F. Effect of mobile telephones on sperm quality: A systematic review and meta-analysis. *Environment International*. 2014;70:106–12.
87. Avendano C, Mata A, Sanchez Sarmiento CA, Doncel GF. Use of laptop computers connected to internet through Wi-Fi decreases human sperm motility and increases sperm DNA fragmentation. *Fertility and Sterility*. 2012;97(1):39–45.e2.
88. Sylvester C, Menke M, Gopalan P. Selective serotonin reuptake inhibitors and fertility: Considerations for couples trying to conceive. *Harvard Review of Psychiatry*. 2019.
89. Franceschi M, Perego L, Cavagnini F, Cattaneo AG, Invitti C, Caviezel F et al. Effects of long-term antiepileptic therapy on the hypothalamic-pituitary axis in man. *Epilepsia*. 1984;25(1):46–52.
90. Cahill DJ, Fox R, Wardle PG, Harlow CR. Multiple follicular development associated with herbal medicine. *Human Reproduction (Oxford, England)*. 1994;9(8):1469–70.
91. Thompson S, editor. Antioxidant supplements, effect on sperm function and the significance of lifestyle factors. *Bristol Fertility Society Annual Conference*; 2019; Liverpool.
92. Cookson J, Hodgson R, Wildgust HJ. Prolactin, hyperprolactinaemia and antipsychotic treatment: A review and lessons for treatment of early psychosis. *Journal of Psychopharmacology (Oxford, England)*. 2012;26(5 Suppl):42–51.
93. Wardle PG, Cahill DJ. *Understanding Infertility*: Editor Smith T. Family Doctor Publications, Dorset, United Kingdom, 2005.
94. Santini SJ, Cordone V, Falone S, Mijit M, Tatone C, Amicarelli F et al. Role of mitochondria in the oxidative stress induced by electromagnetic fields: Focus on reproductive systems. *Oxidative Medicine and Cellular Longevity*. 2018;2018:5076271.
95. Duffy DM. Novel contraceptive targets to inhibit ovulation: The prostaglandin E2 pathway. *Human Reproduction Update*. 2015;21(5):652–70.
96. McInerney KA, Hatch EE, Wesselink AK, Rothman KJ, Mikkelsen EM, Wise LA. Preconception use of pain-relievers and time-to-pregnancy: A prospective cohort study. *Human Reproduction (Oxford, England)*. 2017;32(1):103–11.
97. Matyas RA, Mumford SL, Schliep KC, Ahrens KA, Sjaarda LA, Perkins NJ et al. Effects of over-the-counter analgesic use on reproductive hormones and ovulation in healthy, premenopausal women. *Human Reproduction (Oxford, England)*. 2015;30(7):1714–23.

Assisted conception 8

OVERVIEW

This chapter describes briefly how assisted conception developed from the early days of 1978 to the present, recognising the efforts made to move this from an experimental treatment to a very complex and successful treatment, offering infertile couples realistic opportunities for having their own baby. The major clinical complications associated with assisted conception are discussed. The basic principles of treatment, some details and perspectives on treatment options and advances which might materially add to the success of assisted conception are addressed. Currently recognised risks and discussions on long-term complications are included as well.

8.1 INTRODUCTION

Assisted conception (AC) or assisted reproductive technology (ART) refers to a range of treatments from the moderately simple intrauterine insemination to the complex *in vitro* fertilisation with blastocyst transfer.

Inability to produce a child has been a feature of human experience since the earliest times. Perhaps the beginning of the modern art/science of reproductive medicine was in the 1820s, when ova were identified in menstruating women, but the ovulatory cycle was not understood for another 100 years. The 1840s was when doctors and scientists understood that pregnancy started with the penetration of the egg by male sperm. The first recorded case of successful artificial insemination in the United States was in 1884. Progesterone, oestrogen and testosterone were discovered in the 1920s, and in the ensuing years, it proved possible to manufacture these hormones for use in treatment. As laboratory knowledge and skill increased, many centres around the world vied to fertilise human eggs *in vitro* and transfer them to the uterus. It was 1978 when the first ever baby was born by *in vitro* fertilisation, the result of the pioneering work of a doctor, Patrick Steptoe, an embryologist, Robert Edwards and a scientist, Jean Purdy. The baby, Louise Brown, and her parents became known worldwide. Forty years later, success rates and techniques have evolved and improved, and over 5 million babies have been born worldwide using *in vitro* fertilisation (IVF) techniques.

Assisted conception techniques or assisted reproductive technologies (the terms are interchangeable) have developed enormously over the past 40 years, and I have been fortunate to work in this field for the past 28 of those years. The innovations which I value the most for their beneficial effects in the development of treatment are: the introduction of GnRH agonists to prevent premature ovulation and reduce cycle cancellation; the switch to using a vaginal as opposed to a laparoscopic approach for oocyte collection and, most recently, the introduction of intracytoplasmic sperm injection. The last technique finally provided a solution for the problem of sperm dysfunction. Laboratory techniques have also improved. Chief among these improvements is the ability to freeze and thaw oocytes and embryos while retaining their viability, in a process called vitrification, and, even more astounding, scientists have discovered how to facilitate the 'growth' of the embryo to the 5- to 6-day-old blastocyst stage, an amazing achievement.

8.2 *IN VITRO* FERTILISATION: IT'S COMPLEX!

There's a famous 1893 English cookery book, *Mrs Beeton's Cookery Book*, which contains a recipe and instructions for rabbit stew. The writer rather charmingly gives the first instruction – 'first catch your rabbit'. It has become something of a catchphrase. Performing IVF is rather like following a recipe where the first instruction is – first catch your egg.

Louise Brown was born using natural-cycle IVF, a moderately inefficient form of treatment, as only one egg (oocyte) is collected per cycle. Natural-cycle IVF relies on the production of endogenous FSH for follicular growth, with endogenous LH triggering oocyte maturation (1). Following Louise Brown, a range of stimulatory regimens was introduced to increase oocyte yield. Initially using clomifene and injected FSH as the gonadotrophin, practice moved to using injectable gonadotrophins only. However, an unwanted LH surge caused a high rate of premature ovulation, which made for many wasted cycles. The introduction of an additional drug to the regimen, a gonadotrophin-releasing hormone analogue, (initially agonists), prevented the unwanted premature LH surge and enabled better oocyte collection (2). This meant that IVF treatment could be planned much more effectively, and within a short time facilitated the widespread rollout of the intervention.

To produce an embryo, one must follow these steps:

- Stimulate the ovary using GnRH-a + FSH/HMG + USS monitoring
- Retrieve oocyte/s (see the illustration in Figure 8.1) – *now you have caught your egg*!
- Inseminate the oocytes in a laboratory
- Grow the embryo in culture for 2–3 days – or more, see blastocysts below
- Transfer the embryo (4–6 cells) or blastocysts to the uterus

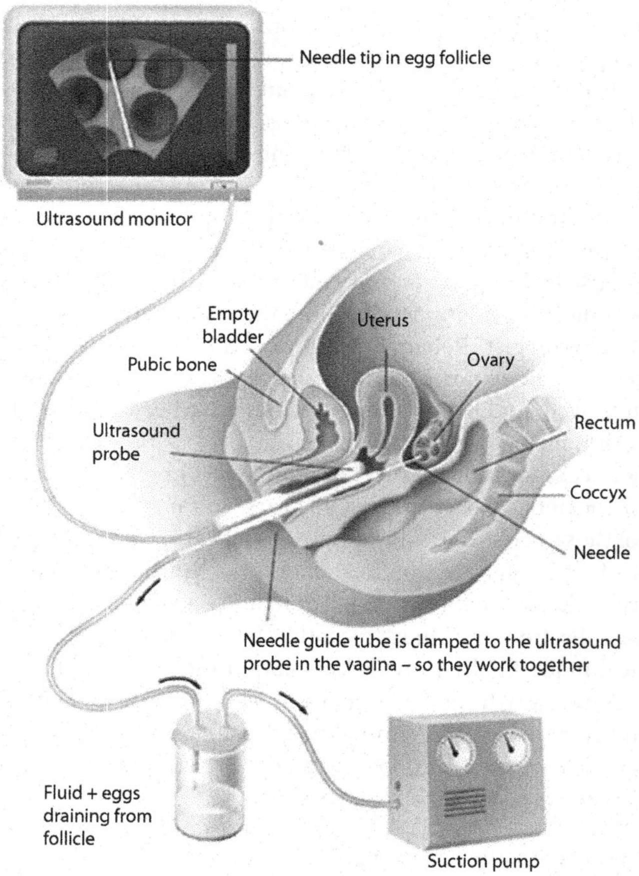

Figure 8.1 The ovary, with distended and mature follicles, can be visualised on ultrasound. The ovary is pierced using a hollow needle and the fluid aspirated into a collecting chamber. This is passed to a scientist waiting to inspect the fluid for an oocyte. (From Wardle PG, Cahill DJ. *Understanding Infertility*: Family Doctor Publications, Dorset, United Kingdom, 2005, with permission.)

Listing each step so briefly risks demeaning the importance and complexity of each individual one. Each step is a masterpiece of clinical or scientific expertise. The clinical skills require time and practice to develop, as individuals learn how to judge the growth rate of follicles and decide that there are adequate mature follicles to harvest them. The individual operator has to be trained to develop a delicacy of touch when undertaking egg collection, entering the follicle to drain it without rupturing it; and she/he must avoid the large internal iliac vein, which has the same proportions as a follicle (about 20 × 20 mm). It usually takes a good trainee a year to develop all the clinical skills which are integral to the process.

Scientific expertise and rigour are required to undertake the laboratory processes. These processes include the grading of oocytes as they are collected. Oocyte grading has both subjective and objective elements to it and is dependent on the appearance of the cumulus cells around the oocyte and how closely packed they are. A mature oocyte is recognised by presence of the first polar body (meaning that meiosis has resumed) and an expanded 'fluffier' cohort of cumulus cells around the oocyte. The culture media required to maintain an embryo are complex, particularly as there is a move to transferring embryos back to the uterus at blastocyst stage, usually 4–5 days into their life. Debate continues as to which is better at supporting embryo and blastocyst culture – it would appear that pregnancy rates are similar whether a single culture system is used or if sequential systems are used, changed once or twice during the culture process (3).

8.2.1 Some recent advances in assisted conception

Techniques to freeze ovarian tissue are still 'developmental' and available to only a few patients, because the number of centres offering this are few and the skills required are extensive. There are hundreds of frozen ovarian tissue samples in laboratories which were gathered using older techniques and which are probably now no longer useful.

Developments now in use or on the immediate horizon include

- The use of GnRH agonist for triggering of ovulation
- Vitrification
- Whole genome analysis

In Chapter 5 (Section 5.3.1), I describe the role of GnRH agonists and antagonists in controlled ovarian stimulation (designed to stimulate multiple follicles and to generate multiple oocytes with the intention of freezing some or all for later use). Antagonist and agonist cycles (both arms used hCG to trigger ovulation) showed no difference in positive pregnancy (34% vs 35%) or live birth rates (22% vs 23.8%) (4), but ovarian hyperstimulation syndrome (OHSS) was an occasional, serious complication of agonist cycles. Using logic arising from a knowledge of the HPA axis and pharmacology, work began in the early 2000s to find a better process, one minimising the occurrence of OHSS. From 2010 onwards, it became clearer and clearer that using a GnRH agonist *instead of* hCG triggered ovulation effectively but produced lower rates of ovarian hyperstimulation. One study reported how the incidence of OHSS fell from 3% when using hCG as trigger to no cases when using a GnRH agonist as trigger (9). An RCT of over 1000 women found a reduction in severe OHSS (5.1% vs 8.9%) and moderate OHSS (10.2% vs 15.6%) in the GnRH antagonist group with an agonist as trigger, compared with the agonist group with hCG as trigger (4,6,8). Severe OHSS (severe morbidity) rates vary in standard IVF protocols using GnRH agonists to trigger ovulation; initially, incidences ranged from 1% to 14% (6); more recent clinical experience and publications report lower OHSS rates – and for severe OHSS on the order of 1% or less (7).

While using GnRH agonists to trigger ovulation diminished the severity and likelihood of OHSS, it had unwanted effects on the luteal phase of the cycle. Without the ongoing drive from hCG, there was early demise of the corpora lutea and impaired endometrial receptivity to the implanting embryo. Supplemental doses of hCG or progesterone were used by the researchers to remedy this. Perhaps it is because of the need for that luteal phase intervention and lack of clarity about optimal protocols to use, but the uptake of GnRH agonists in GnRH antagonist down regulated cycles is low, with clinics using it between 5% and 36% of cases (5). Given its apparent risk reduction, moving to use GnRH agonists to trigger ovulation in GnRH antagonist downregulated cycles would seem to benefit patients.

Vitrification of oocytes and embryos was introduced in the mid-2000s and has been gradually adopted in most centres, replacing the previous method of slow freezing. Vitrification solidifies the sample – oocyte or embryo – into a glasslike state, avoiding the formation of both intra- and extracellular ice. The process uses high concentrations of cryoprotectant agent and very high cooling rates (15,000–30,000°C/min). Vitrification does not involve expensive instrumentation, so it is cheaper, and because the process is so fast, it is more time efficient, requiring only a few minutes as compared with 1–2 hours with standard freezing (10). It provides relatively high survival rates for embryos/blastocysts, of the order of 80%. It seems likely that this will have an increasing role in embryo preservation.

Whole genome analysis and preimplantation genetic diagnosis. There is a maxim in medicine that you should only undertake an investigation if you think a positive result will change your management of the patient. Whole-genome analysis of a couple's DNA and preimplantation genetic diagnosis (PGD), in my opinion, comes into this category. The entire chromosomal content of an individual or embryo is screened, and this allows the identification of obscure or maybe unimportant but abnormal genes. If the screened entity is an embryo, the abnormality is managed by not replacing that embryo. It was hoped, when it was introduced, that preimplantation genetic diagnosis of the embryo/blastocyst would provide information about common issues such as recurrent miscarriage and implantation problems. A 2019 European multicentre study showed that in women over 35 years, the use of PGD for aneuploidy screening of the first and second polar body did not increase the live birth rate (11).

Some men and women are infertile because of relatively obscure chromosomal causes. Identifying that cause using PGD gives a certain amount of satisfaction, but the abnormality is not (at present at any rate) amenable to treatment, and the procedure does not improve a couple's chances of getting pregnant and having a baby. All this may mean that PGD falls out of favour – time will tell.

8.3 WHAT IS APPROPRIATE WHEN?

Assisted conception or assisted reproduction treatments range from intrauterine insemination (usually with ovulation induction) to *in vitro* fertilisation and embryo transfer to *in vitro* fertilisation by intracytoplasmic injection and embryo transfer. In addition, there is the option of longer culture of embryos to the blastocyst stage. Complexity and cost of treatment increases across the range.

The reasons a couple might end up looking at assisted conception as a treatment option will vary, of course, but most people will have one of the diagnoses in the box, in line with the diagnostic categories outlined in Chapters 2 and 3. IVF and/or ICSI are the last endeavour or even the last resort of couples who want to get pregnant and who have exhausted all other options. For all the diagnoses listed in the box, good outcomes at IVF would be expected (12). This is even more the case now that ICSI means that sperm disorders are no longer associated with poor outcomes at IVF as clinical pregnancy rates with ICSI are as good as those as with IVF.

Indications for assisted conception

Tubal disease
Ovulatory dysfunction
Sperm dysfunction (mild/moderate)
Endometriosis
Unexplained
Failed other treatment options

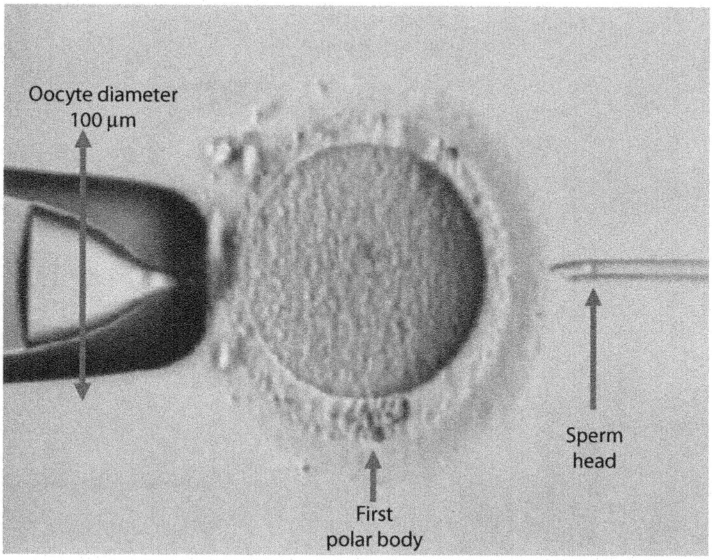

Figure 8.2 Intracytoplasmic sperm injection: The injecting needle on the right, with the sperm head arrowed, the holding pipette on the left and the oocyte diameter (100 μm).

Oocytes and embryos differ in their behaviour when being cryopreserved. Oocytes are single-celled structures and are generally in a state of preparation for meiosis, especially if they have been exposed to hCG. The image of the oocyte in Figure 8.2 (as it is about to undergo ICSI) helps to illustrate some of what is going on inside the oocyte. The polar body on the outside is matched on the inside by the nuclear structure, with spindles attached to each chromosome acting like a delicate brittle skeleton inside the cell. That cytoskeleton is easily damaged, and if that were to occur, the next stages of meiotic activity, preparing for the arrival of a sperm and its chromosomes, will not function. Embryos and blastocysts are different because they have multiple cells, and they are not always in a state of nuclear mitotic activity. Consequently, they are more robust and resilient when frozen – less fragile and more likely to survive the process.

8.4 THE RANGE OF TREATMENT OPTIONS

8.4.1 Intrauterine insemination

Intrauterine insemination is tangentially addressed in Chapter 5 (Section 5.4). This approach to treatment has been in use for many years. Its origins are hazy, but people in the late nineteenth century in the United States (James Marion Sims) and Europe (Ilya Ivanovich Ivanoff) are usually credited with its introduction (13).

Technical advances included the use of prepared sperm and injected gonadotrophins. Prepared sperm use didn't really start until the era of Steptoe and Edwards, when prepared sperm was required for IVF. Before that, unprepared sperm were injected directly into the uterus, sometimes causing uterine contractions and infections in the upper genital tract. Sperm are prepared by removing prostaglandins, infectious agents, antigenic proteins, leucocytes and immature germ cells and then inseminated into the uterus 2–3 hours after preparation.

IUI is undertaken in cycles stimulated by gonadotrophins, with ovulation usually triggered by an injection of hCG (analogous to LH). Ovulation needs to be triggered for IUI because the maturation process of the oocyte must be completed to allow meiosis to begin and for the oocyte to extrude its first polar body in preparation for oocyte penetration by a sperm. For IUI, the woman is awake, placed in lithotomy position, with the cervix being seen using a speculum; sperm are inseminated by direct transcervical passage of a fine plastic catheter into the uterine cavity (as illustrated in Figure 8.3).

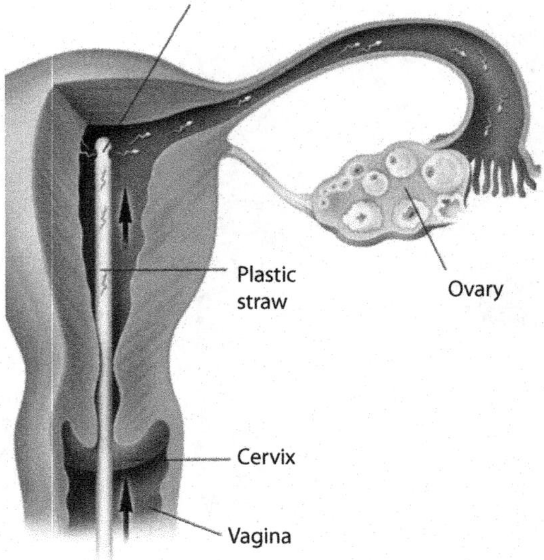

Figure 8.3 Intrauterine insemination. (From Wardle PG, Cahill DJ. *Understanding Infertility*: Family Doctor Publications, Dorset, United Kingdom, 2005, with permission.)

The indications for intrauterine insemination include unexplained infertility, minor endometriosis, ovulatory disorders (including PCOS) and mild sperm dysfunction. Adding IUI in to the treatment of these conditions serves to increase the success rate of treatment of ovulation induction. This means that relatively simple technology can be used to treat more people. Intrauterine insemination requires both patent and healthy tubes and 'reasonable' quality sperm. Reasonable in this sense will vary slightly from laboratory to laboratory, but the laboratory we have worked with uses an indicator, the total motile sperm count (TMS), derived after sperm have been processed through a 'recovery' process, and the laboratory suggests that the TMS should exceed 1 million motile sperm in order to be able to give a reasonable chance of oocyte fertilisation after insemination (14). The results of recovery and survival are predictive of success at IVF, but not at IUI. The markers in IVF (oocyte fertilisability and pregnancy rates) are useful as surrogate markers for IUI, which is why they are used in this way – those data, of course, predate ICSI (see Section 8.4.3). Published data on IUI success rates from the UK assisted conception regulatory authority, the HFEA, give a figure of 14% for women under 35 and 4% for the over 40s (18). Other authors give higher figures – 20% (16) and 24% (17). The IUI service where I worked achieved yearly figures of ranging from 9% to 20% (see Table 5.2 in Chapter 5). Remember, all these figures are for IUI with ovulation induction.

The complications of treatment are multiple pregnancy, ovarian hyperstimulation treatment and using the wrong sperm for insemination. Multiple pregnancies can be minimised by not allowing IUI to go ahead when more than two mature follicles are present. Ovarian hyperstimulation syndrome has been greatly reduced by not allowing triggering of ovulation with hCG when they are too many follicles – also serving to minimise multiple pregnancy. The last complication, using the wrong sperm, should be a 'never' event, something spoken of but prevented completely (in a world where no human error exists) by quality control systems.

8.4.2 *In vitro* fertilisation and embryo transfer

The first baby born as a result of IVF was Louise Brown. Steptoe and Edwards, and Jean Purdy, an important and often forgotten colleague (19), had worked together for many years on this challenge, from the late 60s onwards until the announcement of the delivery of Louise Brown (20). There had been a considerable amount of competition with colleagues in the United States (21) and in Australia (22) all working on this. In a sense, the British team was lucky to get there first.

IVF is an excellent treatment for almost all the indications in the box above, with the exception of sperm dysfunction. It effectively treats all causes of infertility, including damaged tubes, and allows opportunities for couples desiring to have a child the opportunity to do so providing they can get access to these facilities. Differing countries have differing levels of access to assisted conception, which vary depending on the country's wealth and disposable income. In the United Kingdom, Europe and North America, success rates are available in published form, usually annually, from the Human Fertilisation and Embryology Authority (HFEA) (18), from the European Society of Human Reproduction and Embryology (23) and from the Society for Assisted Reproductive Technology (24). Data from the HFEA for 2016 show pregnancy rates (as live birth rates) above 20% on average (with a wide variation by age – 15% for women over 40, 29% for women under 35 years of age) (25). Data will always be a year or two behind, as it takes time to assimilate and then examine the data from each centre.

With IVF, the complications of treatment are multiple pregnancy, ovarian hyperstimulation treatment (OHSS), complications arising from surgery (infection and bleeding) and (as for IUI) using the wrong sperm for insemination. As Figure 8.1 illustrates, the technique of oocyte/egg collection involves the passage of a needle (no more than 2 mm diameter) with a tip which is visible on ultrasound. This usually requires some form of sedation for the patient. The needle is passed into every follicle and the fluid (containing the egg) is drained. Fluid from the follicles is aspirated and passed to the lab staff to identify eggs. The needle can enter other organs in the pelvis, notably blood vessels (the internal iliac vein is closest) and the small bowel. Occasionally, a small endometriotic or blood-filled cyst will be entered if not recognised as such. Multiple pregnancy rates have fallen dramatically over 30 years – in 1991, multiple pregnancy rates from fresh embryos were 30%, while in 2016, they were 11% (25). Efforts being made to reduce the impact of multiple pregnancy are addressed in Section 8.5.5. The effects of multiple pregnancies are covered in Section 8.7.2. OHSS rates have fallen similarly and are covered later in Section 8.7. Surgical complications are now rare – recognition of the error is key. Infection is rare and bleeding is reported to occur in 0.25% of cases (26). Withdrawing frank blood probably implies the needle is in the internal iliac vein, and needle withdrawal with pressure on the vein for 60 seconds will usually stop the bleeding. In my experience over many years, only one encounter in a woman (who had a previous undiagnosed blood clotting disorder) resulted in major haemorrhage (26). Draining an endometriotic cyst should be covered by broad-spectrum antibiotics.

8.4.3 Intracytoplasmic sperm injection

Intracytoplasmic sperm injection was introduced into clinical practice in 1992; the person credited with its introduction is Andre van Steirteghem from Brussels (27). For couples in which the man had a moderate to severe sperm dysfunction, 'ordinary' IVF was ineffective with very low success rates until the advent of ICSI (27). Prior to the announcement of the success of ICSI, several other techniques were tried – partial zona dissection (28) and subzonal insemination (29). ICSI has revolutionised the treatment of men with severe sperm disorders, giving the possibility of pregnancy where that was not possible, and has radically altered for the better the treatment of men with poor sperm quality.

For couples undergoing ICSI as opposed to IVF, there is really no difference as far as they are concerned. It isn't until the eggs and sperm are in the laboratory that things are different. The technique involves the collection of one single sperm in a fine needle, first removing its tail (to ablate any motility and prevent it swimming back out), then piercing the egg far away from the first polar body (extruded from the egg when the first meiotic division happens). In this way, the chromosomes and spindles attached to them are least likely to be damaged by the process (see Figure 8.2).

For the ICSI technique, a special mount for the microscope is used which absorbs all vibration away from the process, as even a passing vehicle could affect it.

This new technique was the first successful method to overcome the difficulty of penetrating oocytes. Up to then, that problem was insurmountable and meant that the diagnosis of a serious sperm disorder was associated with very low fertilisation rates and consequently low chances of pregnancy. ICSI, once introduced, spread widely as a technique. The monitoring of pregnancies arising from ICSI as a treatment was much more intensive than it was when IVF was introduced in 1978. The results from that monitoring have been generally reassuring. There were no differences in neurodevelopmental scores or other mental developmental indices, though birthweights were lower in children born after ICSI (30) and data from Western Australian studies suggested a higher rate of birth defects (twofold) compared to natural conception (31). These ICSI-conceived children (boys only) were more likely than naturally conceived children to need corrective surgery for genitourinary conditions, 5% vs 1% (32). In a relatively recent review of abnormalities associated with assisted conception, the authors suggest that the likelihood of abnormalities associated with ICSI are increased, but not significantly (33). Their article provides some interesting insights into why the incidence may be falling – it may be related to a dilutional effect of more men having ICSI for less rigorous reasons than when it was first introduced, and it's also possible that improved techniques have also reduced the incidence. There is very little to add on the complications associated with ICSI, as the operative and laboratory procedures are as those for IVF apart from the long-term consequences discussed above.

ICSI has increased in use since its introduction – to about 36% of cycles in 2016 and live births from ICSI in that year in the United Kingdom were 7500, about 37% of the total live births (25). In the United States, about 66% of cycles used ICSI, but the data do not provide for live birth rates from assisted conception to ICSI alone (34). However, birth rates are 30% for those under 35 years of age and 11% for those over 40 years of age (34).

8.4.4 Cryopreservation and embryo transfer

From the 1980s onwards, embryo laboratories have looked at methods to more usefully use the embryos left over in a treatment cycle, after the two, three or more embryos have been transferred. One obvious thing to do was to cryopreserve them with a view to thawing at a later stage. Initially this was not very successful, and very few pregnancies were achieved with thawed embryos. As time went on, freezing techniques were altered and improved. Freezing biologically active tissue will vary in difficulty depending on the tissue. Sperm are relatively easy to freeze, as the cytoplasm-to-nucleus ratio is very low and the cell is not actively dividing. The sperm samples, after screening for infections, are mixed with cryoprotectant, cooled slowly and plunged into liquid nitrogen. Oocytes and embryos differ in their behaviour when being cryopreserved. Oocytes are single-celled structures and are usually in a state of preparation for meiosis, especially if they have been exposed to hCG to prompt ovulation and oocyte recovery. The image of the oocyte in Figure 8.2 (as it is about to undergo ICSI) helps to illustrate some of what is going on inside the oocyte. The polar body on the outside is matched on the inside by the nuclear structure, with spindles attached to each chromosome acting like a delicate brittle skeleton inside the cell. That cytoskeleton is easily damaged, and if that were to occur, the next stages of meiotic activity, preparing for the arrival of a sperm and its chromosomes, will not function properly.

Embryos and blastocysts are different because they have multiple cells, and they are not always in a state of nuclear mitotic activity. Consequently, they are more robust and resilient when being frozen – and more likely to survive the process. However, like the oocyte, but not as fragile, the cell spindle structures within the embryo are easily damaged if the fluid within the cell freezes too quickly. As ways of drawing out the intracellular fluid have improved, and especially since 'vitrification' has been introduced, success rates in embryo freezing/thawing have increased – with success rates above 20% on average (25).

8.5 ADDITIONAL ELEMENTS TO THE PROCESS OF *IN VITRO* FERTILISATION AND EMBRYO TRANSFER

8.5.1 Preimplantation diagnosis

Preimplantation diagnosis was first introduced as a technique in 1990. The present indications for its use are screening for chromosome abnormalities, HLA typing and monogenic diseases, preimplantation genetic screening (PGS) and social sex selection. It is not widely practiced, as the range of expertise to deliver this service is extensive – embryo biopsy, embryo culture and rapid examination of the chromosomal content usually using fluorescent *in situ* hybridisation (FISH). Most recent data on the outcomes in Europe are available (35). Only a few centres in the United Kingdom are licensed to do PGD (36). In Europe, PGD has been introduced, but ethical and moral questions led to its initial banning in Italy and Latvia and long delays in its introduction in France because of legislation clashes (37). In the Gulf States, of 60 IVF clinics, 4 offered PGD in 2009. The low numbers were attributed to cost, the technology involved (which has no other use in clinical practice) and poor returns on investment (38). Uptake in the United States seems to be quite different. Of the 493 clinics surveyed in the United States, 449 (91%) offered access to PGD (39).

8.5.2 Endometrial scratching

The scientific investigation of implantation has focussed interest on a number of candidate genes which might promote and support implantation, and it was noted that mechanical disruption of the endometrium led to an increase in expression of proimplantation factors – these include proteins such as integrins and metalloproteinases (40). Barash and team showed that endometrial injury in the natural cycle prior to IVF increased the chance of pregnancy twofold in live birth after endometrial scratches compared to control (41). In a 2014 randomised controlled trial of 300 women, there was no significant difference in ongoing pregnancy rates in the treated group compared to the untreated group [26.7% (40/150) vs 32.0% (48/150), respectively; RR 0.833 (95% CI 0.585–1.187) (42)]. A later Cochrane review identified eight eligible studies in couples having IUI or natural intercourse (43). All of these were graded as being either 'low quality evidence' or 'very low-quality evidence' and required interpretation with caution. They concluded that endometrial injury may improve clinical pregnancy rates with an RR of 1.98 (95% CI 1.51 to 2.58). A further randomised controlled trial was published in 2019 (44). In this multicentre study of 1364 women, well powered to ensure the results are meaningful, the frequency of live birth after IVF in the endometrial-scratch group was 26.1%, and in the control group, it was 26.1%. The associated editorial in *New England Journal of Medicine* (NEJM) spells out the message from this paper: 'Scratching the Endometrium in *In Vitro* Fertilization—Time to Stop' (45).

8.5.3 Assisted hatching

When an oocyte is released at ovulation, it is surrounded by a dense group of cells which are derived from the cumulus oophorus and which must be penetrated by sperm to achieve fertilisation – these cells and the oocyte itself secrete a dense protective layer made up of glycoproteins, known as the zona pellucida. Once the oocyte is fertilised, this layer surrounds the embryo and effectively contains it. The zona pellucida must be disrupted and the growing embryo (now a blastocyst) allowed to 'hatch' at 5–6 days of age to be free to grow further and then to implant, maybe 1 day later. Spontaneous hatching appears dependent on a number of genes whose expression is correlated with hatching. These genes produce cathepsin V, GATA-binding protein 3 and human chorionic gonadotropin beta subunit 3, 5, 7 and 8. Hatching does not appear to be dependent on zona pellucida or gamete quality (46). Artificial hatching of the oocyte was first described in 1991 by Cohen and his group (47). It relies on breaching of the zona pellucida by chemical (acidified Tyrode's solution), by laser or by mechanical methods (using a glass microneedle) (48). The ASRM paper comprehensively covers the risks and complications as well as the value of assisted hatching – in essence, to date, there is a small (though significant) increase in pregnancy rates (and multiple pregnancy rates) associated with the use of assisted hatching (49).

8.5.4 Blastocyst transfer

An *embryo* is the term used in assisted conception to describe a 2–8-cell structure growing after fertilisation; then it is called a *morula*, and a *blastocyst* describes the stage of development at 5–6 days of age. By the blastocyst stage, differentiation into the inner cell mass (the cells that will become the fetus) and the trophoblast (which will become the placenta) has occurred. Data from animal studies gave indications that allowing embryos to be cultured for several days longer would ensure that implantation was more likely to occur when transferred. Culture systems need to be able to cope with the changing nutritional demands of the growing embryo, and much effort has gone into identifying the best culture mediums and sequence. The risk with longer culture times is that the embryos fail to grow, and so there may be no embryos available for transfer, but the corollary is that even if transferred at 2–3 days, the implantation rate is not good, and pregnancy may not ensue. Research in humans found that transferring blastocysts compared with 3–4 cell embryos gives higher pregnancy rates, but only in women over 35; the benefit was not seen in younger women (48% vs 20%, $p = 0.016$) (50). The following year, in 2016, a Cochrane review published findings based on 27 RCTs studying cleavage-stage versus blastocyst-stage embryo transfers. The conclusion based on low-quality evidence was that fresh blastocyst transfer showed slightly better pregnancy rates (36%) than cleavage-stage embryo transfer (29%) per couple (9 RCTs; OR 1.35, 95% CI 1.05 to 1.74) (83). The review authors were unimpressed with the quality of the studies and advocated further better-designed research to be able to advise patients properly.

8.5.5 Single embryo transfer

When IVF was introduced, it was not uncommon to replace all the embryos that were produced. This was partly because implantation rates were low, and chances of success were greater the more embryos were transferred. It was also because any extra embryos would be wasted, as techniques for embryo freezing were not developed. Patients as well as the doctors thought it worth the risk of a multiple pregnancy. As pregnancy success rates improved, multiple pregnancy rates rose in tandem. By the end of the twentieth century, opinion swung to limiting the number of fetuses transferred *in utero* because of instances of large multiple births. Medical bodies and regulators around the world were quick to recognise the poor and negative message that this provided to the public. Campaigns such 'One at a time' and others called for reductions in embryo numbers transferred (51). The Human Fertilisation and Embryology Authority in the United Kingdom strongly advise that only one embryo/blastocyst be replaced at any one time. Fortunately, the freezing process has improved with time so that embryos can be saved for a later attempt if necessary, and the usual practice now is to transfer a single embryo at a time.

In Japan, the single embryo transfer (SET) policy was successful in its objective in that it led to a decrease in twin pregnancies (from 33.9% to 13%) and in neonatal intensive care unit admissions (52). However, predictably, the live birth rate per cycle also fell. In 1200 women randomised to either single or double embryo transfer, live birth rates were lower after single (27%) compared to double embryo transfer (42%) (53). A key element of single embryo transfer is a commitment to replace excess embryos in further cycles. When this is added in to the data, live birth rates from SET and one frozen cycle were 38%, not much different to the live birth rates obtained after one fresh double embryo transfer (42%) (53).

The pressures for guidance to minimise the chances of multiple birth are sensible and laudable, but as with many decisions we make, they have an effect on someone. In my clinic's most successful year for pregnancy following ovarian stimulation and IUI, the pregnancy rate was 29%, but that came at a 'cost' of four multiple pregnancies, two twins and two triplets. We reviewed those cases and recognised that they were all young women who had three or more follicles, with serum oestradiol levels in excess of 2000 pmol/L (equivalent to <600 pg/mL). Changing our policy in the light of

UK HFEA recommendations reduced the multiple pregnancy rate, but the overall pregnancy rate also fell, to under 20%. Fertility treatments are expensive and a couple's resources often limited. Understandably, they want the best chance of success in every cycle.

Bizarrely, even in the best regulated of circumstances, when one or two embryos are replaced, multiple pregnancies can occur – in one case, quadruplets after a two-embryo transfer (54). The quadruplet pregnancy in that case appeared to be related to intercourse the night before oocyte collection, leading to advice to couples to abstain from unprotected intercourse in treatment cycles.

8.5.6 Donated gametes

For a couple, the use of donated gametes requires a significant shift in expectation and an acceptance of the loss of biological parenthood by the member of a couple who cannot provide the needed gametes. Women who need to use donated eggs will be drawn from patient groups who have had an early menopause (either natural, surgically induced or medically induced, usually by chemotherapy) or women who have had unexpected failure of their oocytes at IVF after all other causes have been ruled out. For men, the indications for the use of donated gametes are broadly the same – failure of sperm production (following surgery damaging the hypothalamus or pituitary or the testis, or following chemotherapy for testicular or other cancers) and failure of fertilisation at IVF despite the use of intracytoplasmic sperm injection. Overcoming hypothalamic or pituitary failure as a cause of spermatogenesis failure is dealt with fully in Chapter 4 (Section 4.6.1). Changes in legislation can affect the availability of options – in the United Kingdom, regulatory changes related to gamete donation meant that donors' identity could be revealed once the resultant child reached 18. This had a negative effect on the number of men willing to be sperm donors. It is interesting that while the United Kingdom experience was a fall in the numbers of donors (55), the experience was different in Australia (56). Why this is the case is unclear – many Australian clinics noted an initial reduction in numbers, but recruitment strategies and media campaigns served to increase donor numbers within a short space of time.

8.5.7 Surrogacy

Surrogacy is the process whereby a couple who wants to have a child (the commissioning couple) enter an agreement with a woman who is willing to carry their prospective child (the surrogate), whereby she gets pregnant, carries their baby and hands it to them soon after delivery. Most surrogacy arrangements use assisted conception techniques to produce an embryo which is then transferred to the surrogate's uterus (gestational surrogacy). A few surrogate pregnancies are based on the use of (usually) the male partner's sperm to inseminate the uterus of the surrogate mother (traditional surrogacy). Because this is usually arranged on a personal basis, with sperm delivered directly, the numbers of couples using this latter format is impossible to know. The arrangements for surrogacy fall into two categories – altruistic or commercial/gainful. Recognition of surrogacy in law differs greatly by country. For instance, it is banned in Italy (57) and restricted in Portugal since 2016 (58). In the United Kingdom, surrogacy is permitted if altruistic (59). However, only a small number of UK assisted conception clinics offer gestational surrogacy. The very small numbers are probably because the entire process is liable to run into difficulties (59). The potential for misunderstanding and conflict is high, the potential for the surrogate not to agree to hand over the child is low in terms of probability but enormous in terms of adverse publicity and the investment in counselling before treatment is considerable. In the United States, the situation varies according to each individual state. Surrogacy is banned by some, permitted but unregulated by others, permitted but limited to altruistic arrangements only or, finally, permitted and commercially regulated. Data is not gathered state by state, but it is known that 2.4% of 660,000 IVF cycles in the United States in 2014 used gestational surrogate carriers (60).

8.6 AT WHAT COST?

The assisted conception process is costly – in terms of finance (whoever is funding the treatment), but also in emotional, physical and relationship terms. Financial costs vary by country and within countries. In the United States, recent costs were estimated at an average of $12,000. In Europe, average costs are £3000 (equivalent to $3900) per cycle. In India, the average costs are Rs. 250,000 (equivalent to $3500).

The pressures that couples feel are beyond 'will it work or not', and concerns about the growth of follicles, the numbers of eggs produced and the quality of the semen sample produced tend to be all internalised and produce major pressures throughout the period of treatment. A further desire to keep the treatment private is hindered by the need to attend clinics for multiple appointments, which makes that all the more stressful. It might not occur to a casual onlooker how much the treatment affects people. The extent of the effect is seen in several markers – including the rate of relationship breakdown, which is worsened when pregnancy doesn't ensue. One French study followed couples for up to 6 years after their initial consultation for fertility problems in their hospital. Of the 550 who responded to follow-up, 53 had separated and in that group; being childless increased the likelihood of separation (28% in that group compared to 4% in those with at least one child) (61). A small local study showed little impact, with relationships failing in only 5% of respondents (71). Provision of counselling for couples is an important element of preparation for and recovery from cycles of assisted conception. Interventions such as cognitive behavioural therapy and psychological support before assisted conception measures were associated with increased pregnancy rates compared to controls: 55% and 54% of those having an intervention experienced a viable pregnancy, compared to 20% of controls (62).

8.7 RISKS AND COMPLICATIONS

- Ovarian hyperstimulation syndrome
- Multiple pregnancy
- Long-term complications for the woman and the offspring

8.7.1 Ovarian hyperstimulation syndrome

The first two of these, ovarian hyperstimulation syndrome and multiple pregnancy, were a significant concern when I started to work in assisted conception in 1990.

Ovarian hyperstimulation syndrome occurs after the triggering of luteinisation by hCG and usually therefore after egg collection. It is primarily a clinically diagnosed condition, characterised by abdominal pain and distension, nausea, vomiting and oliguria, with ultrasound evidence of ovarian swelling and biochemical evidence of haemoconcentration. Twenty-five years ago, OHSS was a relatively common complication, occurring in 2%–5% of pregnancies following controlled ovarian stimulation and hCG triggering, with one admission every month to my hospital ward with relatively severe OHSS. This is not a benign condition, with an incidence of up to 10% deep vein thrombosis (63,64) and occasional mortality (65).

Since the 1980s and 90s, strategies to prevent OHSS have been developed, and it is now a very rare event. Principal strategies in prevention have been the introduction of:

- 'Coasting' – withholding gonadotrophins until serum oestradiol concentrations and follicular size begin to fall
- Withholding hCG and the use instead of biological LH to trigger ovulation
- The opportunity to 'freeze all' embryos and withholding hCG to support the luteal phase and instead using a natural progesterone formulation (available by pessary, suppository or injection)
- The use of a GnRH agonist to trigger ovulation in cycles controlled by GnRH antagonists (see Section 8.2.1)

Strategies and guidelines for the management of OHSS are published by several recognised specialist societies: British Fertility Society (66), the Royal College of Obstetrics and Gynaecology (67), the European Society for Human Reproduction and the American Society for Reproductive Medicine (68).

The major pathologies associated with ovarian hyperstimulation syndrome are ovarian distension, a fluid shift from the intravascular space to the extravascular space giving rise to ascites and haemoconcentration, and pain and nausea. Management focuses on control of fluid balance, drainage of fluid collections, thrombosis prevention and, rarely, the need for intensive care in the event of a severely ill woman. Fluid replacement uses the oral route if tolerated, with intravenous fluids such as colloids rather than crystalloids, which will not stay in the intravascular space. Draining of ascites would be undertaken if this were causing severe discomfort and/or respiratory difficulties. The use of thromboprophylaxis should be universal for women admitted with OHSS and in the rare eventuality that clotting disorders and fluid management are difficult to control, then admission to an intensive care unit is advised (66).

8.7.2 Multiple pregnancy

According to the UK Office of National Statistics, pregnancy following assisted conception is 11 times more likely to result in a multiple birth, and in 2014, 16% of pregnancies from assisted conception resulted in a multiple birth. Multiple pregnancies occur when more than one embryo is replaced after IVF or when ovulation induction therapy produces several mature eggs, with or without intrauterine insemination. Multiple pregnancies are an undesired outcome because of the risks to the fetus and the mother and because of the high demand they place on clinical resources. In addition, multiple pregnancies can cause problems for the parents' emotional health, relationship and finances.

The management of multiple pregnancies and their complications is a specialised clinical area, generating research, conferences and textbooks dedicated to the topic. Even the least complicated multiple pregnancy, dizygotic twins, carries more risk of complications and poor outcome than singleton pregnancies. All multiple pregnancies have a high risk of preterm birth (OR 1.69, 95% CI 1.62–1.77), caesarean delivery (RR 1.15, 95% CI 1.13–1.17) and pre-eclampsia (OR 1.17, 95% CI 1.11–1.22) (69). Low gestational age at delivery is associated with poorer outcomes, either increased mortality or long-term morbidity – bronchopulmonary dysplasia and chronic lung disease, severe visual impairment resulting from retinopathy of prematurity, hearing impairment, cerebral palsy and cognitive developmental delay. Figure 8.4 provides an illustrative summary of the data from the EPICure study, showing mortality associated with premature delivery and consequently expectations for survival (70).

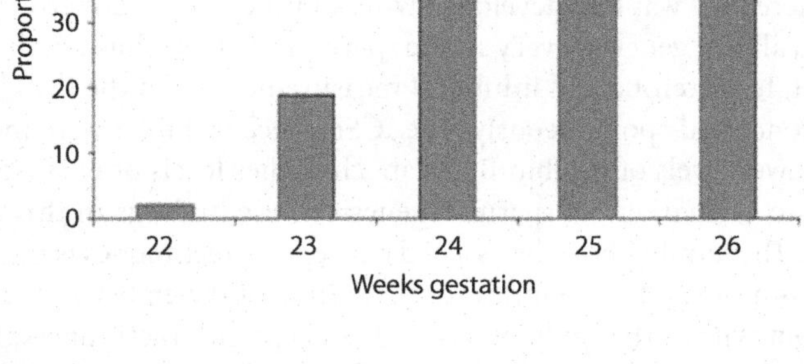

Figure 8.4 Live born survival rates expressed as proportions at increasing gestational age, weeks. (Data from Costeloe KL et al. *BMJ (Clinical Research Ed)*. 2012;345:e7976.)

The impact of these multiple pregnancies causes a considerable drain on neonatal services – for example, a quadruplet pregnancy delivering at 34 weeks 15 years ago led to cumulative 154 days of care in the neonatal unit (54).

In addition to the possible physical consequences for the baby and the mother, multiple pregnancies put the relationship of the couple under strain. I recall one sad story of a woman who had been successful at IVF coming back 2 or 3 years later with her triplet babies, delivered after treatment. They had done well in terms of medical complications, but the father of the children had left the family home. But anecdotes and impressions spark research, and the research findings on the effect of multiple pregnancies on relationships are inconclusive. There are difficulties conducting such research. Large-scale data, for example, census reports, cannot accurately capture the breakdown of long-term cohabiting and less formal relationships, so that the main outcome reported is divorce, which is more easily quantified, is less and less informative and generalisable as rates of marriage decline. Also, engaging in fertility treatments is in itself stressful, and this can be confounding when examining data on what happens to couples after the baby/babies are born.

Data from the US census of 1980 is useful in respect to the impact of multiple births, because it predates the widespread use of IVF, and the twins it followed were (presumably) naturally conceived. It did not, however, report on higher multiples than twins (72). Divorce rates between those with twins and those without did not differ significantly overall. However, divorce rates of those with twins in the first pregnancy compared with the whole cohort did show slight but significant levels of difference, (OR 1.08, 95% CI 1.01–1.16; predicted absolute risk 13.7% vs 12.7%, $p = 0.02$). These are old data (pre-1980), and the subject we are considering, multiple pregnancies against a background of infertility, is not similar to the cohort population captured in that census. The UK Twins and Multiple Births Association report on the effect of multiple births on families draws on data from the United Kingdom–based Millennium Cohort Study and Family Resources Survey (73). Based on their findings, they concluded that families with twins or higher multiples were inclined to fracture more than those of singletons, and the results were statistically significant. More importantly, the report details the financial and material distress experienced by the whole family when multiple children arrive at once.

8.7.3 Long-term complications for the offspring and the woman

Long-term complications include an increase in abnormalities in children born from assisted conception, and particularly intracytoplasmic sperm injection. Even in singleton pregnancies, when assisted conception is compared to spontaneously conceived conceptions, there appears to be an increased risk of preterm delivery, low birth weight and the need for intensive care (74). Singletons born to infertile treated couples have a higher rate of congenital malformations and (with less strength of association) of rare imprinting disorders (conditions such as Beckwith–Wiedemann Syndrome and Angelman's Syndrome) (74). Initial reports on neurodevelopmental outcome were not reassuring, but longer-term studies show no difference in developmental attainment between children born after assisted conception and after natural conception (74). Children born as a result of ICSI are, in 2018, moving into adulthood, and continue to be the subject of scientific scrutiny. The Brussels unit where ISCI was first developed will examine the reproductive ability of these men, fathered by men who all had generally very severe sperm disorders. Thus far, they have taken serum samples for FSH, LH, testosterone and inhibin B measurement from the ISCI offspring group and controls who were conceived spontaneously. The ICSI-conceived men were more likely than controls to have much lower levels of inhibin B and much higher levels of FSH (75). As inhibin B and FSH are recognised to be markers of spermatogenesis, these findings in this preliminary dataset are clearly worrying. The number born by assisted conception techniques is increasing year by year, and with that, there is more and more innovative handling of the embryo such as growing it to the blastocyst stage. From within the body of assisted conception practitioners, there is emerging a desire to employ more stringent monitoring of the children born by these means, and a caution of forging ahead too quickly without extensive modelling and research in animals. Tessa Roseboom

Table 8.1 Key positive findings with respect to assisted conception treatment.

Risk factor	Relative risk[a]
Cancer of the ovary	
Never conceiving from ART	1.67 (1.42–1.95)
Having 1–2 cycles of ART	1.38 (1.10–1.70)
Risk of ovarian cancer by any female factor	1.66 (1.46–1.88)
by endometriosis	2.47 (1.75–3.39)
by tubal damage	1.71 (1.40–2.08)
Cancer of the breast	
In situ breast cancer	1.15 (1.02–1.29)
Invasive breast cancer	0.96 (0.92–1.00)

Source: Data from Williams CL et al. *BMJ (Clinical Research Ed)*. 2018;362:k2644.

[a] This relative risk compares the 3155 of the cohort who had some form of cancer with the 252,631 who did not.

made these points cogently and comprehensively in her 2018 paper (76), while Bart Fauser and colleagues argued for the need for a follow-up programme for these children (84). Mulder and his colleagues, on the other hand, have called for more extensive animal model testing before these techniques are used extensively in humans (85).

Although concerns about risks to the woman have been hopefully put to rest by a report in 2018 (77), in the past there has been serious concern about the increased risk of ovarian cancer in infertile women being treated with drugs to improve ovulation (78). Over time, attention focussed on just clomifene, and on further investigation the link was recognised as minimal or nonexistent (79). An examination of 113,226 women who had assisted conception, followed up for a mean of 4.87 years, found no increase in incidence of all cancers compared with standardised rates (80). The 2018 review by Williams and colleagues examined data held by the Human Fertilisation and Embryology Authority (225,786 women who underwent assisted reproduction in Great Britain, 1991–2010) and also examined this question, specifically focussing on cancers of the endometrium, breast and ovary (77). Key positive findings from this paper are presented in Table 8.1. It is important to note that ovarian tumour risk was not increased in women who were treated because of male-only or unexplained infertility. The increased risk was in women with endometriosis, low parity or both, with low parity being a well-established risk factor for ovarian cancer. There was no increased risk of endometrial or invasive breast cancer. These data are very reassuring for our patients.

8.8 CASE STUDY

Fatima and Sherif were married in their early 20s and 5 years later began to try for a family. Fatima was then 27. Within 6 months, Fatima became anxious about having a child and went to see her family doctor. The doctor explained that they would not be referred to a specialist, as the couple had only been trying for 6 months. Fatima broke down and cried at the appointment, emphasising the distress she was feeling and the pressure she felt to have a child. The doctor told her she could be referred, but as she had been trying for only 6 months, the health authority would not pay for investigations or treatment and she and Sherif would have to pay to see someone.

The couple decided to pay to see a specialist, and they were seen within a month. The specialist asked lots of questions, examined both of them, and arranged many blood tests. Six weeks later, they returned for the results. The doctor went through all the tests and told Fatima and Sherif that nothing wrong had been found so far. One couldn't really label this as infertility and certainly not unexplained infertility, as the duration of time was insufficient. They were pleased at this, but then realised that no cause meant no obvious pathway to treatment. The doctor outlined the possibilities in terms of actual treatments to get pregnant. These included intrauterine insemination and *in vitro* fertilisation, all having different possibilities for success. It was very confusing.

They left, a little downhearted and uncertain. The doctor had provided them with some information to take away and, after some long talks together, they decided to go back to talk about *in vitro* fertilisation. The specialist approved the idea of IVF, said they could go to a private IVF clinic, and provided details about costs. The costs were high, but they had enough savings to afford it.

They went through the first cycle, which Fatima found very stressful. Her ovaries responded poorly despite daily injections of a gonadotrophin (150 IU FSH daily), and she felt it was all her fault. However, they managed to collect six good-quality eggs. Sherif's sperm were added to the eggs that day, and they waited for the call telling them about the fertilisation. Two days later, one of the nurses phoned them to say the embryo transfer would be that afternoon, but sadly, only one of the six eggs had fertilised. This was duly transferred, but her period came 12 days after the transfer.

At the next appointment, the doctor talked about the low number of eggs collected and the poor fertilisation of those eggs. She suggested trying again, but this time using a higher dose of injections, 300 units of FSH drug to begin with. She proposed that when the eggs were collected, the scientists in the laboratory carry out a special technique to soften and weaken the shell around the egg (assisted hatching). Fatima thought this sounded very far fetched, but she trusted the doctor and agreed to go ahead. Eight weeks later, into her second cycle, Fatima was responding better to the medications and at egg collection, 12 eggs were collected. Of those, 10 were fertilised and 2 replaced after assisted hatching had been done. This again did not lead to a pregnancy.

8.8.1 Commentary

For whatever reasons, Fatima and Sherif wanted to find out why they had not conceived after 6 months of trying. When no obvious cause was found, they agreed to move to *in vitro* fertilisation quite quickly. Some couples reach a point in their lives when they want a child now, and they can be impatient with advice to keep trying and wait. In addition, privately funded patients often don't adhere to standard advice.

The first cycle, quite unexpectedly, led to a poor ovarian response, the subject of intense research over the years. In a woman of this age, with normal ovarian reserve (which was considered to be the case on the basis of normal day 5 FSH levels), a poor response is most likely to be due to chance.

But patients often want to see a change in plan, and adding in more gonadotrophins is reasonable under the circumstances. More open to question is the suggestion of assisted hatching. This procedure is of uncertain value and often reserved for women over 35 or 38 years of age. There is little value in assisting hatching in someone of Fatima's age. Replacing two good embryos in a young woman is probably also unwise. One randomised study examined clinical pregnancy rates in women who had one and two embryos replaced – there was no difference in pregnancy rates, and 11 of 39 women in the two-embryo transfer group had twin pregnancies, with one multiple (monozygous) pregnancy in the single-embryo transfer group (81); note this differs from McLernon et al. (53).

Think about what advice you might offer Fatima in her third cycle. The hardest and bravest option is to do nothing – but that inertia will not satisfy some patients.

8.9 SUMMARY

At present, assisted conception or artificial reproductive technologies comprise oocyte collection, sperm recovery, intrauterine insemination, *in vitro* fertilisation embryo transfer and intracytoplasmic sperm injection. Advances in clinical and laboratory procedures have increased the chances of successful conception by improved freezing of oocytes and embryos, and of successful implantation by improved and longer culture of the embryo. Two of the serious complications of these procedures – ovarian hyperstimulation and multiple pregnancies – have been almost eliminated by changes in clinical practice, but congenital abnormalities in the offspring from ICSI remain a major cause of concern. Recently introduced techniques – assisted hatching, endometrial scratching, whole genome analysis and preimplantation genetic diagnosis – are still under scrutiny although offered by some centres. Above all, there is need for close scrutiny of the health and functioning of children born by these artificial means, especially as the embryo is subject to more and more manipulation and interference.

REFERENCES

1. Fahy UM, Cahill DJ, Wardle PG, Hull MG. *In-vitro* fertilization in completely natural cycles. *Human Reproduction (Oxford, England).* 1995;10(3):572–5.
2. Hughes EG, Fedorkow DM, Daya S, Sagle MA, Van de Koppel P, Collins JA. The routine use of gonadotropin-releasing hormone agonists prior to *in vitro* fertilization and gamete intrafallopian transfer: A meta-analysis of randomized controlled trials. *Fertility and Sterility.* 1992;58(5):888–96.
3. Dieamant F, Petersen CG, Mauri AL, Comar V, Mattila M, Vagnini LD et al. Single versus sequential culture medium: Which is better at improving ongoing pregnancy rates? A systematic review and meta-analysis. *JBRA Assisted Reproduction.* 2017;21(3):240–6.
4. Toftager M, Bogstad J, Bryndorf T, Lossl K, Roskaer J, Holland T et al. Risk of severe ovarian hyperstimulation syndrome in GnRH antagonist versus GnRH agonist protocol: RCT including 1050 first IVF/ICSI cycles. *Human Reproduction (Oxford, England).* 2016;31(6):1253–64.
5. Lambalk CB, Banga FR, Huirne JA, Toftager M, Pinborg A, Homburg R et al. GnRH antagonist versus long agonist protocols in IVF: A systematic review and meta-analysis accounting for patient type. *Human Reproduction Update (Oxford, England).* 2017;23(5):560–79.
6. Mathur RS, Joels LA, Jenkins JM. Ovarian hyperstimulation syndrome may be more likely if multiple pregnancy occurs following assisted conception. *Acta Geneticae Medicae et Gemellologiae.* 1995;44(3–4):233–5.
7. Al-Ramahi M. Severe OHSS: Decreasing the risk of severe ovarian hyperstimulation syndrome. *Human Reproduction (Oxford, England).* 1999;14(10):2421–2.
8. Iliodromiti S, Blockeel C, Tremellen KP, Fleming R, Tournaye H, Humaidan P et al. Consistent high clinical pregnancy rates and low ovarian hyperstimulation syndrome rates in high-risk patients after GnRH agonist triggering and modified luteal support: A retrospective multicentre study. *Human Reproduction (Oxford, England).* 2013;28(9):2529–36.
9. Alyasin A, Mehdinejadiani S, Ghasemi M. GnRH agonist trigger versus hCG trigger in GnRH antagonist in IVF/ICSI cycles: A review article. *International Journal of Reproductive Biomedicine (Yazd, Iran).* 2016;14(9):557–66.
10. AbdelHafez FF, Desai N, Abou-Setta AM, Falcone T, Goldfarb J. Slow freezing, vitrification and ultra-rapid freezing of human embryos: A systematic review and meta-analysis. *Reproductive Biomedicine Online.* 2010;20(2):209–22.
11. Verpoest W, Staessen C, Bossuyt PM, Goossens V, Altarescu G, Bonduelle M et al. Preimplantation genetic testing for aneuploidy by microarray analysis of polar bodies in advanced maternal age: A randomized clinical trial. *Human Reproduction (Oxford, England).* 2018;33(9):1767–76.
12. Hull MG. Infertility treatment: Relative effectiveness of conventional and assisted conception methods. *Human Reproduction (Oxford, England).* 1992;7(6):785–96.
13. Ombelet W, Van Robays J. Artificial insemination history: Hurdles and milestones. *Facts, Views & Vision in ObGyn.* 2015;7(2):137–43.
14. Ford WC, Mathur RS, Hull MG. Intrauterine insemination: Is it an effective treatment for male factor infertility? *Bailliere's Clinical Obstetrics and Gynaecology.* 1997;11(4):691–710.
15. Stovall DW, Guzick DS, Berga SL, Krasnow JS, Zeleznik AJ. Sperm recovery and survival: Two tests that predict *in vitro* fertilization outcome. *Fertility and Sterility.* 1994;62(6):1244–9.
16. Mills MS, Eddowes HA, Cahill DJ, Fahy UM, Abuzeid MI, McDermott A et al. A prospective controlled study of *in-vitro* fertilization, gamete intra-fallopian transfer and intrauterine insemination combined with superovulation. *Human Reproduction (Oxford, England).* 1992;7(4):490–4.

17. Nandi A, Bhide P, Hooper R, Gudi A, Shah A, Khan K et al. Intrauterine insemination with gonadotropin stimulation or *in vitro* fertilization for the treatment of unexplained subfertility: A randomized controlled trial. *Fertility and Sterility.* 2017;107(6):1329–35.e2.
18. HFEA. *Fertility treatment 2014. Trends and figures [PDF file].* London: HFEA; 2016 [cited 2017 15/12/2017]. Available from: http://ifqtesting.blob.core.windows.net/umbraco-website/1783/fertility-treatment-2014-trends-and-figures.pdf.
19. Edwards RG, Steptoe PC, Purdy JM. Fertilization and cleavage *in vitro* of preovulator human oocytes. *Nature.* 1970;227(5265):1307–9.
20. Steptoe PC, Edwards RG. Birth after the reimplantation of a human embryo. *Lancet (London, England).* 1978;2(8085):366.
21. Garcia J, Acosta A, Andrews MC, Jones GS, Jones HW, Jr., Mantzavinos T et al. *In vitro* fertilization in Norfolk, Virginia, 1980–1983. *Journal of In Vitro Fertilization and Embryo Transfer: IVF.* 1984;1(1):24–8.
22. Lopata A, McMaster R, McBain JC, Johnston WI. *In-vitro* fertilization of preovulatory human eggs. *Journal of Reproduction and Fertility.* 1978;52(2):339–42.
23. Calhaz-Jorge C, De Geyter C, Kupka MS, de Mouzon J, Erb K, Mocanu E et al. Assisted reproductive technology in Europe, 2013: Results generated from European registers by ESHRE. *Human Reproduction (Oxford, England).* 2017;32(10):1957–73.
24. SART. *ART Success Rates.* Washington, DC: CDC; 2015. Available from: https://www.cdc.gov/art/artdata/index.html.
25. HFEA. *Fertility Treatment 2014–2016 – Trends and Figures.* HFEA; 2016. Available from: https://www.hfea.gov.uk/media/2563/hfea-fertility-trends-and-figures-2017-v2.pdf.
26. Cahill DJ, Fox R, Wardle PG. Ureteral obstruction – A complication of oocyte retrieval. *Fertility and Sterility.* 1994;61(4):787–8.
27. Palermo G, Joris H, Devroey P, Van Steirteghem AC. Pregnancies after intracytoplasmic injection of single spermatozoon into an oocyte. *Lancet (London, England).* 1992;340(8810):17–8.
28. Cohen J, Malter H, Fehilly C, Wright G, Elsner C, Kort H et al. Implantation of embryos after partial opening of oocyte zona pellucida to facilitate sperm penetration. *Lancet (London, England).* 1988;2(8603):162.
29. Fishel S, Antinori S, Jackson P, Johnson J, Rinaldi L. Presentation of six pregnancies established by sub-zonal insemination (SUZI). *Human Reproduction (Oxford, England).* 1991;6(1):124–30.
30. Sutcliffe AG, Taylor B, Saunders K, Thornton S, Lieberman BA, Grudzinskas JG. Outcome in the second year of life after *in-vitro* fertilisation by intracytoplasmic sperm injection: A UK case-control study. *Lancet (London, England).* 2001;357(9274):2080–4.
31. Hansen M, Kurinczuk JJ, Bower C, Webb S. The risk of major birth defects after intracytoplasmic sperm injection and *in vitro* fertilization. *New England Journal of Medicine.* 2002;346(10):725–30.
32. Bonduelle M, Wennerholm UB, Loft A, Tarlatzis BC, Peters C, Henriet S et al. A multi-centre cohort study of the physical health of 5-year-old children conceived after intracytoplasmic sperm injection, *in vitro* fertilization and natural conception. *Human Reproduction (Oxford, England).* 2005;20(2):413–9.
33. Pinborg A, Henningsen AK, Malchau SS, Loft A. Congenital anomalies after assisted reproductive technology. *Fertility and Sterility.* 2013;99(2):327–32.
34. SART. *ART Success Rates.* CDC; 2015 [cited 2015]. 'ART Success Rates'. CDC, accessed 15/12/2017. https://www.cdc.gov/art/artdata/index.html. Available from: https://www.cdc.gov/art/artdata/index.html.
35. De Rycke M, Belva F, Goossens V, Moutou C, SenGupta SB, Traeger-Synodinos J et al. ESHRE PGD Consortium data collection XIII: Cycles from January to December 2010 with pregnancy follow-up to October 2011. *Human Reproduction (Oxford, England).* 2015;30(8):1763–89.

36. Genetic Alliance UK. *How Can I Access Preimplantation Genetic Diagnosis?* Genetic Alliance UK; 2016 [cited 2017 27.11.2017]. Available from: http://www.geneticalliance.org.uk/information/services-and-testing/how-can-i-access-preimplantation-genetic-diagnosis/.
37. Duguet AM, Boyer-Beviere B. Preimplantation genetic diagnosis: The situation in France and in other European countries. *European Journal of Health Law.* 2017;24(2):160–74.
38. Eskandarani HA. Pre-implantation genetic diagnosis in the Gulf Cooperative Council countries: Utilization and ethical attitudes. *Human Reproduction and Genetic Ethics.* 2009;15(2):68–74.
39. Capelouto SM, Archer SR, Morris JR, Kawwass JF, Hipp HS. Sex selection for non-medical indications: A survey of current pre-implantation genetic screening practices among U.S. ART clinics. *Journal of Assisted Reproduction and Genetics.* 2018;35(3):409–16.
40. Goel S, Sharma R, Hamilton A, Beith J. LHRH agonists for adjuvant therapy of early breast cancer in premenopausal women. *Cochrane Database of Systematic Reviews.* 2009(4):Cd004562.
41. Barash A, Dekel N, Fieldust S, Segal I, Schechtman E, Granot I. Local injury to the endometrium doubles the incidence of successful pregnancies in patients undergoing *in vitro* fertilization. *Fertility and Sterility.* 2003;79(6):1317–22.
42. Yeung TW, Chai J, Li RH, Lee VC, Ho PC, Ng EH. The effect of endometrial injury on ongoing pregnancy rate in unselected subfertile women undergoing *in vitro* fertilization: A randomized controlled trial. *Human Reproduction (Oxford, England).* 2014;29(11):2474–81.
43. Lensen SF, Manders M, Nastri CO, Gibreel A, Martins WP, Templer GE et al. Endometrial injury for pregnancy following sexual intercourse or intrauterine insemination. *Cochrane Database of Systematic Reviews.* 2016(6):Cd011424.
44. Lensen S, Osavlyuk D, Armstrong S, Stadelmann C, Hennes A, Napier E et al. a randomized trial of endometrial scratching before *in vitro* fertilization. *New England Journal of Medicine.* 2019;380(4):325–34.
45. Mol BW, Barnhart KT. Scratching the endometrium in *in vitro* fertilization – Time to stop. *New England Journal of Medicine.* 2019;380(4):391–2.
46. al-Nuaim LA, Jenkins JM. Assisted hatching in assisted reproduction. *BJOG: An International Journal of Obstetrics and Gynaecology.* 2002;109(8):856–62.
47. Cohen J. Assisted hatching of human embryos. *Journal of In Vitro Fertilization and Embryo Transfer: IVF.* 1991;8(4):179–90.
48. ASRM. Role of assisted hatching in *in vitro* fertilization: A guideline. *Fertility and Sterility.* 2014;102(2):348–51.
49. Li D, Yang DL, An J, Jiao J, Zhou YM, Wu QJ et al. Effect of assisted hatching on pregnancy outcomes: A systematic review and meta-analysis of randomized controlled trials. *Scientific Reports.* 2016;6:31228.
50. Fernandez-Shaw S, Cercas R, Brana C, Villas C, Pons I. Ongoing and cumulative pregnancy rate after cleavage-stage versus blastocyst-stage embryo transfer using vitrification for cryopreservation: Impact of age on the results. *Journal of Assisted Reproduction and Genetics.* 2015;32(2):177–84.
51. Hamberger L, Hardarson T. Multiple pregnancy – The evidence for single-embryo transfer. *Human Fertility.* 2003;6(sup2):S65–S7.
52. Hayashi M, Satoh S, Matsuda Y, Nakai A. The effect of single embryo transfer on perinatal outcomes in Japan. *International Journal of Medical Sciences.* 2015;12(1):57–62.
53. McLernon DJ, Harrild K, Bergh C, Davies MJ, de Neubourg D, Dumoulin JCM et al. Clinical effectiveness of elective single versus double embryo transfer: Meta-analysis of individual patient data from randomised trials. *BMJ (Clinical Research Ed).* 2010;341:c6945.
54. Cahill DJ, Jenkins JM, Soothill PW, Whitelaw A, Wardle PG. Quadruplet pregnancy following transfer of two embryos: Case report. *Human Reproduction (Oxford, England).* 2003;18(2):441–3.

55. Tomlinson MJ, Pooley K, Pierce A, Hopkisson JF. Sperm donor recruitment within an NHS fertility service since the removal of anonymity. *Human Fertility (Cambridge, England).* 2010;13(3):159–67.
56. Adams DH, Ullah S, de Lacey S. Does the removal of anonymity reduce sperm donors in Australia? *Journal of Law and Medicine.* 2016;27:144–54.
57. Iliadou M. Surrogacy and the European Court of Human Rights: Reflections on *Paradiso and Campanelli v Italy. Medical Law Review.* 2018; 27:144–54
58. Raposo VL. The new Portuguese law on surrogacy – The story of how a promising law does not really regulate surrogacy arrangements. *JBRA Assisted Reproduction.* 2017;21(3):230–9.
59. Horsey K. Fraying at the edges: UK Surrogacy Law in 2015. *Medical Law Review.* 2016;24(4):608–21.
60. Perkins KM, Boulet SL, Levine AD, Jamieson DJ, Kissin DM. Differences in the utilization of gestational surrogacy between states in the U.S. *Reproductive Biomedicine & Society Online.* 2018;5:1–4.
61. Ferreira M, Sanchez ET, Gatimel N, Fajau C, Lesourd F, Bujan L et al. Parenthood and separation in couples 6 years after their first infertility consultation. *European Journal of Obstetrics, Gynecology, and Reproductive Biology.* 2016;198:7–11.
62. Domar AD, Clapp D, Slawsby EA, Dusek J, Kessel B, Freizinger M. Impact of group psychological interventions on pregnancy rates in infertile women. *Fertility and Sterility.* 2000;73(4):805–11.
63. Mills MS, Eddowes HA, Fox R, Wardle PG. Subclavian vein thrombosis: A late complication of ovarian hyperstimulation syndrome. *Human Reproduction (Oxford, England).* 1992;7(3): 370–1.
64. Kasum M, Danolic D, Oreskovic S, Jezek D, Beketic-Oreskovic L, Pekez M. Thrombosis following ovarian hyperstimulation syndrome. *Gynecological Endocrinology: Official Journal of the International Society of Gynecological Endocrinology.* 2014;30(11):764–8.
65. Cluroe AD, Synek BJ. A fatal case of ovarian hyperstimulation syndrome with cerebral infarction. *Pathology.* 1995;27(4):344–6.
66. Tan BK, Mathur R. Management of ovarian hyperstimulation syndrome guidelines. Produced on behalf of the BFS Policy and Practice Committee. *Human Fertility (Cambridge, England).* 2013;16(3):160–1.
67. Royal College of Obstetricians and Gynaecologists. *The Management of Ovarian Hyperstimulation Syndrome. The Management of Ovarian Hyperstimulation Syndrome.* London: RCOG; 2016. p. 22.
68. ASRM. Prevention and management of ovarian hyperstimulation syndrome. *Fertility and Sterility.* 2016;106(7):1634–47.
69. Oberg AS, VanderWeele TJ, Almqvist C, Hernandez-Diaz S. Pregnancy complications following fertility treatment-disentangling the role of multiple gestation. *International Journal of Epidemiology.* 2018.
70. Costeloe KL, Hennessy EM, Haider S, Stacey F, Marlow N, Draper ES. Short term outcomes after extreme preterm birth in England: Comparison of two birth cohorts in 1995 and 2006 (the EPICure studies). *BMJ (Clinical Research Ed).* 2012;345:e7976.
71. Corrigan E, Cahill DJ, Meadowcroft J, Hull MGR. Social outcome of couples following assisted conception treatment. *Annual ESHRE Conference*; 1999; Tours, France: Hum Reprod 1999;14(suppl 1):47.
72. Jena AB, Goldman DP, Joyce G. Association between the birth of twins and parental divorce. *Obstetrics and Gynecology.* 2011;117(4):892–7.
73. McKay S. *The Effects of Twins and Multiple Births on Families and Their Living Standard.* Twins and Multiple Births Association; 2010.
74. Basatemur E, Sutcliffe A. Follow-up of children born after ART. *Placenta.* 2008;29(Suppl B): 135–40.

75. Belva F, Roelants M, De Schepper J, Van Steirteghem A, Tournaye H, Bonduelle M. Reproductive hormones of ICSI-conceived young adult men: The first results. *Human Reproduction (Oxford, England).* 2017;32(2):439–46.
76. Roseboom TJ. Developmental plasticity and its relevance to assisted human reproduction. *Human Reproduction (Oxford, England).* 2018;33(4):546–52.
77. Williams CL, Jones ME, Swerdlow AJ, Botting BJ, Davies MC, Jacobs I et al. Risks of ovarian, breast, and corpus uteri cancer in women treated with assisted reproductive technology in Great Britain, 1991–2010: Data linkage study including 2.2 million person years of observation. *BMJ (Clinical Research Ed).* 2018;362:k2644.
78. Rossing MA, Daling JR, Weiss NS, Moore DE, Self SG. Ovarian tumors in a cohort of infertile women. *New England Journal of Medicine.* 1994;331(12):771–6.
79. Rossing MA, Tang M-TC, Flagg EW, Weiss LK, Wicklund KG. A case-control study of ovarian cancer in relation to infertility and the use of ovulation-inducing drugs. *American Journal of Epidemiology.* 2004;160(11):1070–8.
80. Luke B, Brown MB, Spector LG, Missmer SA, Leach RE, Williams M et al. Cancer in women after assisted reproductive technology. *Fertility and Sterility.* 2015;104(5):1218–26.
81. Martikainen H, the Finnish ETSG, Tiitinen A, the Finnish ETSG, Tomás C, the Finnish ETSG et al. One versus two embryo transfer after IVF and ICSI: A randomized study. *Human Reproduction.* 2001;16(9):1900–3.
82. Wardle PG, Cahill DJ. *Understanding Infertility*: Family Doctor Publications. Dorset United Kingdom, 2005.
83. Blake DA, Farquhar CM, Johnson N, Proctor M. Cleavage stage versus blastocyst stage embryo transfer in assisted conception. *Cochrane Database Of Systematic Reviews.* 2007(4):Cd002118.
84. Fauser BC, Devroey P, Diedrich K, Balaban B, Bonduelle M, Delemarre-van de Waal HA et al. Health outcomes of children born after IVF/ICSI: a review of current expert opinion and literature. *Reproductive Biomedicine Online.* 2014;28(2):162–82.
84. Mulder CL, Serrano JB, Catsburg LAE, Roseboom TJ, Repping S, van Pelt AMM. A practical blueprint to systematically study life-long health consequences of novel medically assisted reproductive treatments. *Human Reproduction (Oxford, England).* 2018;33(5):784–92.

9 Preserving fertility

OVERVIEW

This chapter describes the options and possibilities for fertility preservation for both medical and social indications. Social preservation of fertility receives much media attention and is generating a lot of interest along economic, sociological and national policy lines. Medical indications have greater time pressures and other difficulties associated with the decision making and are the focus of most clinical effort and research.

9.1 MEDICAL INDICATIONS FOR PRESERVING FERTILITY

Women and men with various medical conditions can preserve their fertility if fertility is threatened by treatment of that condition. My first encounter with this issue was meeting young women presenting with early stages of breast cancer in their late 20s and early 30s. They faced the possibility that chemotherapy could destroy their follicles, wiping out any chance of having children should they survive the cancer. I have been involved with fertility preservation efforts in cases like this ever since. I have also obtained sperm from men who were unable to ejaculate because of different conditions – paraplegia, multiple sclerosis and neuropathies. Preserving the future fertility prospects of young teenagers tends to fall under the responsibility of paediatric gynaecologists and adolescent paediatricians. The technology of freezing and storing gametes and embryos is improving rapidly, and the last 10 years have seen a significant step forward with vitrification, a new rapid freezing technique. Now there are credible prospects for being able to offer most women and men a realistic opportunity to preserve fertility if they need to.

The medical conditions which pose a serious threat to future fertility are conditions which need toxic drugs as treatment – cancers, bone marrow transplants and severe rheumatoid arthritis. In addition, there are the conditions which make it impossible to have sexual intercourse, for example, paraplegia and other neurological conditions resulting in failure to ejaculate.

The challenges in preserving fertility for women, men and adolescents differ, and they will be dealt with in this chapter.

9.2 WOMEN, MEN, ADOLESCENTS

Numerically, the most common condition which creates the need for fertility preservation in women of childbearing age is breast cancer, but any kind of cancer could put a woman in this position, depending on the treatment. Sickle cell anaemia which needs bone marrow transplantation is another condition, as is rheumatoid arthritis if the recommended treatment is an alkylating agent. Generally, these women present having just been given a potentially life-changing diagnosis a few days earlier or told that they require treatment which will affect their future fertility. There will be some degree of pressure to start treatment as soon as possible.

Cancer constitutes the most common diagnosis presenting for fertility preservation. The most common cancers in women of childbearing age are breast cancers, which are approximately 5 times more common than cervical and skin cancers, and nearly 10 times more common than brain and ovarian cancers, these five malignant conditions being the most common cancers in women of this age group (www.cancerresearchuk.org).

9.2.1 Treatment effects on fertility

The effects of chemotherapy on fertility for any cancer will depend on two factors: the cytotoxicity of the therapy (see Table 9.1) and the ovarian reserve of the woman. The ovarian reserve will depend on the woman's age and measurable variables such as FSH, anti-Mullerian hormone and the antral follicle count (1).

Table 9.1 Expected effect of chemotherapeutic drugs on fertility on common tumours in childhood and adolescence.

Condition	Drug doses likely to be used (total)	Risk in men	Risk in women
T-cell non-Hodgkin lymphoma	Cyclophosphamide (2–4 g/m^2)	Low	Low (2)
B-cell non-Hodgkin lymphoma Stage 1–4	Cyclophosphamide (3–6.8 g/m^2)	Low to adverse, dose dependent	Low (2,3)
Acute lymphoblastic leukaemia	Cyclophosphamide (2–5.8 g/m^2)	Low to high, dose dependent	Low (4,5)
Several trials using differing doses	Cytosine arabinoside (0.6–24.8 g/m^2)		
Cranial radiotherapy		Low	Low
Craniospinal radiotherapy		Moderate	Moderate
Gonadal radiotherapy		High	High
Acute myeloid leukaemia	Cytosine arabinoside (10.6–35 g/m^2)	Low	Low (6)
Later studies using higher doses haven't got fertility data			
Sarcomas – Osteogenic	Cisplatin (120–480 mg/m^2)	Low	Low (2,3)
	Adding in ifosfamide (56–72 g/m^2)	Moderate	Moderate
Sarcomas – Ewing's	Ifosfamide (60–102 g/m^2)	Adverse	Uncertain
	Busulphan (600 mg/m^2)	High	High (2,7)
Hodgkin's disease	Ifosfamide (60–102 g/m^2)	Adverse	Low (2,8)
	Chlorambucil (168–366 mg/m^2)	Moderate	Low
	Procarbazine (2.8–5.6 g/m^2)	Moderate	Moderate
	Melphalan (200 mg/m^2)	High	Moderate
Neuroblastoma	Cyclophosphamide (2–6 g/m^2)	Low	Low (2,7)
	Busulphan (600 mg/m^2)	High	High
	Melphalan (up to 200 mg/m^2)	High	Moderate
	Abdominal radiotherapy		High (site dependent)
Wilms' tumour (low–high risk)	Cyclophosphamide (2–8 g/m^2)	Low to moderate	Low (2,9)
Relapse	Melphalan (200 mg/m^2)	High	Moderate
	Abdominal radiotherapy	Low	high

Note: Risk assessment cohorts: Low (<10% subfertility), moderate (10%–50%), adverse (50%–90%), high (>90%).

As women age, there is a steadily decreasing pool of oocytes which means that proportionally, chemotherapy or radiotherapy will have a more damaging effect. The type of treatment used and the area it is targeting will have differential effects. Radiotherapy is more toxic to oocyte numbers than chemotherapy (10). Radiotherapy has obvious direct effects if the targeted tumour is near the hypothalamus or the pituitary, or if it is in the lower abdomen. In the case of the ovary, it is estimated that the radiation dose which destroys 50% of human oocytes (LD50) is no greater than 4 Grays (11). The same group more recently suggested that the LD50 could be even lower – perhaps no more than 2 Grays (12), which is highly important given that cancers are usually treated with between 12 and 30 Grays.

Of the chemotherapeutic agents, alkylating agents are the most damaging cytotoxic drugs, followed by the platinum agents and then the taxanes. Antimetabolites and anthracyclines are the least damaging of the chemotherapeutic drugs (10).

9.2.2 Provision of fertility preservation measures

The challenges and obstacles to delivering fertility preservation measures to adolescent and adult women of reproductive age diagnosed with cancer are considerable, and provision of comprehensive integrated services, even in developed countries, is patchy. These challenges are well described in a paper from a well-functioning fertility preservation service in a major cancer centre (13). Particular key issues to resolve were ensuring communication between oncologists and fertility specialists and the ability to fast-track referrals, putting together a team to deliver care, and conducting consultations as effectively as possible to ensure patients can assimilate the knowledge and make decisions.

There is a short timeframe to work within when cancer is diagnosed, with a focus on the part of patient and oncologist to start chemotherapy as soon as possible, and fertility-sparing measures, even when fully discussed and agreed upon, have their own time requirement to complete. While oncologists may impart the information that chemotherapy will affect ovarian and testicular function, often they do not feel adequately equipped to conduct a discussion about the implications for the person's future fertility, nor do they have the information to inform about the fertility preservation options available, or the chances for success. They may struggle to find access to fertility specialists to refer the patient on to for the advice and counselling needed, and they may also be concerned about the costs to patients of any treatments required (13,14). Rapid referral to a reproductive medicine specialist is desirable to be able to address the many queries that crop up. The approach, in the unit I worked in, was to see women within 5 working days. Surprisingly and distressingly, a 2017 survey involving breast cancer oncologists attending two international conferences found that one-third of the 273 respondents had never read international guidelines on fertility and pregnancy issues associated with young women with breast cancer, and nearly one-fifth did not know what fertility preservations options were available to them in their own country (15). The majority (>80%) were familiar with the use of GnHRa to suppress ovarian function as protection during chemotherapy, but up to 50% showed lack of knowledge about other important and key aspects of the management of fertility and pregnancy in premenopausal women with breast cancer. Their findings appear little different from a 2012 survey of 344 American oncologists, where, although 77% thought it was their responsibility to discuss 'fertility issues' with young women with a cancer diagnosis, roughly 40% did not have confidence that they knew the infertility risks with the chemotherapy they prescribed, and only 25% knew about cryopreservation of oocytes (16).

9.2.3 Options for fertility preservation

The available options to preserve fertility are: suppressing ovarian function during chemotherapy using a GnRHa analogue; stimulation of ovaries, oocyte collection, *in vitro* fertilisation and embryo freezing; stimulation of ovaries, oocyte collection and oocyte freezing; and preservation of ovarian tissue by freezing unstimulated ovarian tissue. Figure 9.1 is an example of an information leaflet given to women with cancer seeking advice about their options in a fertility clinic.

9.2.3.1 GnRH agonists to suppress ovarian function

When prepubertal children undergoing chemotherapy were followed up to the age of 15, girls were found to have sustained less gonadal damage than boys (17). This prompted the hypothesis hat a prepubertal state was beneficial in preserving oocyte function. Efforts were made to find ways to induce such a state in postpubertal females, hoping to limit cytotoxic damaging effects on the ovary and oocytes. The GNRH agonists were tested for this purpose because they 'downregulate' ovarian function by suppressing FSH release, and thereby oestradiol and other oestrogens (18). The flaw in the theory is that in the woman undergoing cancer treatment, it is primordial and primary follicles

Women with cancer who desire preservation of fertility

Setting	Division of Women and Children's Services / Reproductive Medicine / IUI clinic
For Staff	Staff within the Reproductive Medicine Department
Patients	Patients with cancer who desire preservation of fertility

The possible range of treatments:

i. Suppression of ovarian function throughout any chemotherapy using a **GnRH analogue**

ii. Stimulation of ovaries, egg collection, *in-vitro* fertilisation and **embryo freezing**

iii. Stimulation of ovaries, egg collection and **egg freezing**

iv. Preservation of ovarian function by **freezing unstimulated ovarian tissue** or follicles.

i. Suppression of ovarian function throughout any chemotherapy using a GnRH analogue (agonist +/- antagonist) In women who have (oestrogen-receptor positive) breast cancer and require chemotherapy, this has been reported to show reasonable success (>80% resuming menses in some studies). The value of this treatment approach is debated and not completely resolved; the most recent analysis of all the published literature suggests that GnRH agonists taken alongside chemotherapy 'for premenopausal women with endocrine-sensitive breast cancer are supportive of clinical benefit'. In addition, in a small subset of studies, comparing the combination of a GnRH agonist with chemotherapy and tamoxifen to chemotherapy alone, there was a reduction in the risk of breast cancer recurrence with the combination treatment. Advice from the Royal College of Obstetricians and Gynaecologists on this approach is guarded, stating that it 'may be useful'.

ii and iii. Stimulation of ovaries, egg collection, in-vitro fertilisation and embryo or egg freezing

Women who request these treatments do so during a very turbulent time of their lives. Concerns about preservation of future fertility are often raised at this time. The desire to do something or anything to preserve fertility makes this a very difficult time. IVF, egg collection and **embryo freezing** is a well-established treatment, with a clear track record and predictable expectations for outcome. On average 8-10 eggs will be collected in a cycle of treatment, and if mixed with sperm, 80% of these will fertilise. All will be frozen in these circumstances and we expect a 20-40% pregnancy rate per cycle from the replacement of such embryos later.

Egg freezing: expectations of outcome are improving (about 15%) compared to embryo freezing, though success rates are improving. In general, a better outcome is likely with embryo freezing than with egg freezing. There is the possibility of more than one subsequent treatment cycle with embryos but perhaps only one treatment cycle in the future with frozen eggs. Couples might therefore wish to go to centres with experience of this.

iv. Preservation of ovarian function by freezing unstimulated ovarian tissue or follicles.
Requests for such a facility should go to one of the several clinics throughout the country that currently provide this treatment option (as research or treatment). Referral can arranged to one of these centres. Report of about 50 pregnancies in the world have thus far been published following this form of treatment.
Centres who offer ovarian tissue freezing:

Edinburgh Assisted Conception Unit, Edinburgh Royal Infirmary; Oxford Fertility unit - 01865 220076 -

http://www.ouh.nhs.uk/patient-guide/leaflets/files/12856Povarian.pdf

Figure 9.1 Information sheet for women with cancer who desire fertility preservation.

which need protection for future use, not those ready to be stimulated by FSH. Primordial and primary follicles are not responsive to FSH or LH, while primordial follicle growth initiation is independent of FSH; therefore, GnRH agonists should not have any effect on them. Nevertheless, there is some evidence, admittedly contested, that the drugs do work (19). The question is – How?

One possibility is that the assumption that primordial and primary follicles are not responsive to FSH or LH is not correct. Roger Gosden and others have shown that while primordial oocytes do not express mRNA for FSH, the very early primary and two-layer follicles have limited capability to express mRNA for FSH, while the later multilaminar follicles do have full capacity to do so (20). They propose that primordial granulosa cells surrounding the oocytes are dependent on growth factors which are themselves secreted by preantral follicles. So, suppressing those preantral follicles would have an upstream effect to prevent any primary and later follicles from maturing. This seems an intellectually satisfying explanation for an observed phenomenon.

The Munhoz meta-analysis examined seven studies of just over 1000 women and found that GnRHa use with chemotherapy led to higher rates of return of menstruation at 6 months (×2.4) and up to 12 months (×1.8) compared with chemotherapy alone (19). GnRHa use is associated with recognised symptoms arising from low oestrogen, but there does not appear to be any negative effect from its use on survival times.

In a separate multicentre study, the Prevention of Early Menopause Study (POEMS) (21), 218 women with hormone receptor negative breast cancer were randomly allocated to GnRHa plus chemotherapy or chemotherapy alone. Those in the GnRHa group had an ovarian failure rate of only 8% compared with 22% in the chemotherapy-alone group. In addition, 23% of the GnRHa group became pregnant in the 5-year follow up compared with 12% in the control group. There were criticisms of this study, for example, a 38% drop out rate (22), but the results are encouraging. The authors comment that the drop-outs were 'mostly due to deaths ($n = 14$) or lack of FSH data' (21). In the final report of the POEM study, there was a greater likelihood of pregnancy (23.1%) in the GnRHa-treated group compared to chemotherapy alone (12.2%) (OR 2.34, 95%CI 1.07–5.11). The report also said that overall survival was nonsignificantly increased in the GnRHa-treated group (HR 0.45, 95%CI 0.19–1.04; $p = 0.06$) (23). In the final report, the ($n = 39$) dropout rate was 15%, lower than the interim, and therefore probably more acceptable, as data which seem to confirm that the addition of a GnRHa to chemotherapy in women of reproductive years is beneficial in attaining a future pregnancy and is likely to prolong survival (23).

9.2.3.2 Freezing oocytes

Techniques for freezing oocytes are regularly changed and improved. All fertility clinics should ensure that they have access to laboratories where best practice is followed. Recent data on freezing and thawing oocytes (illustrated in Figure 9.2a) suggest survival of thawed oocytes at 95% for women <36 years and 85% for women ≥36 years. Women aged 34, 37 or 42 years, each with 20 frozen mature oocytes, can expect to have a 90%, 75% and 37% likelihood of having at least one live birth, respectively (illustrated in Figure 9.2b) (24).

Oocyte freezing is less complex on ethical grounds – the oocytes are those of the woman (patient) only, and should the woman not survive, disposal of the oocytes causes far less moral or ethical difficulty than if embryos were frozen.

For both oocyte and embryo production, the woman will generally require some degree of mild or more significant stimulation by gonadotrophins, resulting in elevated levels of oestradiol (up to 10 times normal). For women whose cancers are potentially oestrogen sensitive (mostly breast cancers), there is some anxiety that those elevated levels may have deleterious effects on the tumour, perhaps promoting growth or spread (25).

To continue the theme of oestrogen receptors a little longer, one study examined differences in ovarian response by whether the women were ER+ or ER– (ER: oestrogen receptor) and advocated the use of tamoxifen or letrozole (a third-generation aromatase inhibitor) along with FSH injections to keep oestradiol levels low. Surprisingly, this did not have a major negative effect on oocyte

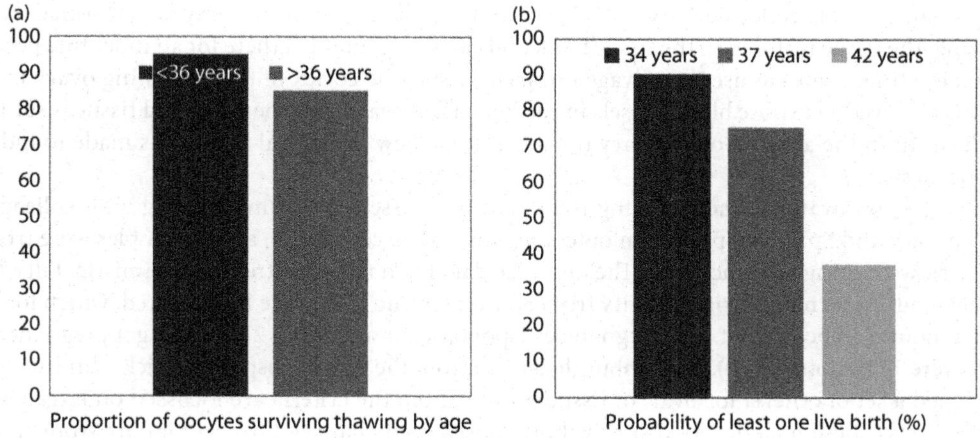

Figure 9.2 (a) Proportion of oocytes surviving thawing by age; (b) the probability of at least one live birth, again by age. (From data in Goldman RH et al. *Human Reproduction (Oxford, England)*. 2017;32(4):853–9.)

outcome, and oocyte numbers were equivalent whether ER+ or ER– (11.5 vs 11.7), and fertilisation rates were 71% vs 60%, respectively (26).

If women and their oncologists express unwillingness to consider ovarian stimulation with gonadotrophins, ovarian tissue freezing remains an option, if available locally, and if the ovaries are intact and healthy.

Having cancer in itself does not appear to be detrimental to the number of oocytes retrieved after controlled ovarian stimulation (for IVF), although this is debated (26). Women who have had chemotherapy with the previous 6 months are likely to have a poorer outcome, probably because the chemotherapy has an effect on the chromosomes or on the spindles holding the chromosomes within the oocyte, making them more fragile, and resulting in fewer good quality oocytes (26).

9.2.3.3 Embryo freezing

Embryo freezing was described and discussed in Chapter 8. The concerns attached to ovarian stimulation in women with breast cancer as already described in *Oocyte freezing* above, apply equally to the process of oocyte collection for the purpose of producing embryos.

9.2.3.4 Freezing unstimulated ovarian tissue

There is ongoing debate as to which method should be used to freeze ovarian tissue, slow freezing or vitrification. When 14 studies were analysed in a meta-analysis, there was no significant difference between slow freezing versus vitrification in terms of the proportion of intact primordial follicles, but vitrification was associated with less primordial follicular DNA strand breaks and better preservation of stromal cells (27). A further study was undertaken in seven healthy volunteers undergoing caesarean section who had ovarian cortex samples collected at the time of their section (28). Each sample was divided in three and was then treated by vitrification, slow freezing and controls which were not frozen (28). Findings from this study related to gene expression: the BAX: BCL-2 ratio was significantly increased in the vitrified group compared to slow frozen and control groups, and slow freezing did not change the susceptibility of the ovarian tissue to apoptotic signals (the BAX: BCL-2 ratio alters the ability of a cell to respond to an apoptotic signal, BCL-2 results in cell protection and BAX promotes apoptosis – and vitrification promotes apoptosis and cell loss) (28). On this and other findings, they concluded slow freezing is optimal for ovary cryopreservation (28). Their views were supported by others after experiments demonstrated that vitrification reduces cell viability (29).

Ovarian tissue is collected laparoscopically and usually an oophorectomy is performed. After freezing, the tissue is stored in the tissue bank and can be maintained there for an indefinite period. When the time comes to use the salvaged ovarian tissue, the cortex of the remaining ovary in the body is removed to expose blood vessels in the bed of the ovary, and the preserved tissue is sutured or glued in. In the absence of an ovary (to attach to), a new peritoneal window is made to induce angiogenesis.

Expertise on ovarian tissue freezing is centred in Brussels, Belgium, working with colleagues in Denmark and Spain. As part of an outcomes study they conducted, all the samples were frozen by the slow-freezing approach (30). The outcome data from these centres are optimistic. Fifty-two women had a return of ovarian activity from 56 women who had tissue reimplanted, which took at least 4 months to be evident. For pregnancies (spontaneous and IVF), 21 women got pregnant and there were 12 live births (30). In Edinburgh, a team from the Royal Hospital for Sick Children have developed a set of criteria for ovarian tissue preservation; the criteria are focussed on largely sensible aspects. These include: good survival prognosis, high chance of ovarian failure from cancer treatment, age less than 30 years, no existing children, no previous radio- or chemo-therapy, and negative HIV and viral hepatitis status (31).

9.2.4 Deciding between fertility-preserving options

The option a woman with cancer will choose is dependent on:

- The type of tumour she has (an oestrogen receptor positive breast tumour might preclude gonadotrophin stimulation for oocyte retrieval); a lymphoma might preclude storing ovarian tissue
- The time span she has before starting chemotherapy/radiotherapy
- Whether she has a male partner (embryo storage is often limited to women with a male partner, as other options are too psychologically complex to deal with in the midst of all the decision making about cancer treatment)
- What is available locally
- Her preference
- What she can afford

Reputable clinics offering preservation of ovarian tissue or oocytes will generally have criteria for selection along these lines:

- The prognosis – this should be good and the expectation from cancer treatment is that it should be curative.
- The preservation of tissue needs to be undertaken within a timeframe not likely to give rise to a delay to the cancer treatment.
- The woman needs to be physically fit enough to undergo a general anaesthetic and a surgical procedure (generally laparoscopic for ovarian tissue preservation or transvaginal for oocyte recovery).
- The patient must be within an age range that is clinically logical (less than 40 years at one extreme, the age of consent at the other extreme).
- Clinics promote embryo storage because the success rates are better, and only if embryo generation has been ruled out will they proceed with oocyte storage.

9.2.5 Fertility preservation for men

Fertility preservation is relatively easy in men, as sperm are much quicker to obtain and easier to freeze than eggs or embryos. Freezing is easier because the nuclear-cytoplasmic ratio is higher in sperm than it is in eggs or embryos; in other words, the amount of cytoplasm is low, making freezing a more successful endeavour. Also, eggs and embryos tend to be in a much more active meiotic or mitotic state than sperm, and consequently the spindles and cytoskeleton in eggs/embryos are much more easily disrupted by freezing. Sperm are relatively easy to obtain by masturbation, no surgery generally being required. For those with neurological pathology, or for men or boys who don't understand how to or who don't wish to masturbate, techniques such as electro-ejaculation or epidydimal aspiration under anaesthesia are likely to provide adequate sperm for storage.

9.2.6 Fertility preservation for children and adolescents

The American Cancer Society lists the following cancers as being the most common in young people: leukaemia, brain and spinal cord tumours, neuroblastoma, Wilms tumour, lymphoma (including both Hodgkin and non-Hodgkin), rhabdomyosarcoma, retinoblastoma and bone cancer (including osteosarcoma and Ewing sarcoma) (32). The prognosis for cancers in childhood and adolescence is constantly improving. Currently, we can expect 80% of children with cancer to be alive 5 years after diagnosis, and up to 70% will be alive and cured at 10 years. Unfortunately, many survivors may have to live with the fertility-damaging consequences of their treatment. It has already been mentioned that the impact of chemotherapy is greater in young boys – follow-up of 240 children treated before puberty with chemotherapy showed that 83% of the boys were azoospermic, while in comparison, 13% of the girls had premature ovarian failure (17).

Freezing ovarian tissue is probably the simplest approach in girls and young women up to the age of 15–16, as ovarian stimulation or suppression is not advisable before or in early puberty (because of the lack of maturity of the HPO axis). It is important to remember that not every cancer is suitable for ovarian tissue freezing; for instance, the haematological cancers can seed secondaries in the ovary.

For young teenage boys whose cancer develops after the onset of puberty, cryopreservation of sperm is technically simple, although for the boy and his parents, the issues are complex in terms of the concept and the impact on his future fertility. Most chemotherapeutic agents are likely to detrimentally affect his spermatogenesis, some more than others. All the alkylating agents (such as cyclophosphamide or chlorambucil) cause azoospermia, as do procarbazine and the platinum-based drugs (33). Scrotal irradiation, without protection of the testicles by a lead sheet, is consistently associated with azoospermia (34).

In one study of 80 boys presenting with a cancer and requesting sperm cryopreservation, 53 were able to easily produce a sample and sperm were cryopreserved. In the rest, the samples contained either no sperm or immotile sperm only. Endocrine markers were generally unhelpful in predicting the acquisition of sperm (35). It is generally recognised that before puberty, and testicular development, attempting sperm recovery after masturbation or electro-ejaculation is unlikely to be unsuccessful, and before the age of 14, it is probably not worth trying (recognising of course that there is biological variation) (35).

Preservation of prepubertal testicular tissue is an area of investigation and research with positive indications that it may be available to all in the future. The results are very promising in animal studies, and human studies are underway. The Brussels group are leading in this – but the treatment is not yet clinically available. Under investigation is how to obtain spermatogonial cells, either from testicular tissue or from testicular cell suspensions.(36) As yet, these techniques are only available as part of a clinical trial (37).

A list of the more common tumours occurring in adolescence, the treatment measures used and the effect on fertility can be found in Table 9.1, using data adapted from a paper by Wallace et al. in Edinburgh (31).

9.3 PREGNANCY AFTER CANCER TREATMENT

As already stated, the use of GnRH agonists offers probable but not guaranteed return of menstruation 6–12 months after cancer therapy (19). The early studies on this showed remarkable preservation of ovarian function, such that at 2 years after the chemotherapy and GnRHa treatment, only 20%–30% of women were still amenorrhoeic, the remainder returning to normal function (38). They also noted however, that with the passage of time, women with normal menstrual function decreased in number (from 70% at 3 years to 50% at 7 years) (38). My approach when reviewing such patients at 2 years or when menstruation returns, is to assess ovarian reserve. The investigations are the same as already described - ultrasound scan to assess antral follicle numbers, assays for FSH, LH and oestradiol, along with anti-Mullerian hormone levels. Evaluation of these results enables the clinician to advise the patient about the urgency of the need to preserve fertility or attempt conception.

Whether using her own frozen eggs or embryos, donor eggs or hoping for a natural conception, the timing of any pregnancy will be a point of discussion between the patient and her oncologist. She will have to weigh her desire to get on with life and start trying for a child against any risks of recurrence caused by the hormone changes which result from a pregnancy. This has been a particular concern with breast cancers over the years, the reasoning being that the high levels of oestrogen hormones throughout pregnancy could act as a stimulus to any occult cancer cells in the breast. There is no consensus on whether a pregnancy following breast cancer is detrimental – several studies, including a large meta-analysis of 14 studies, found that, surprisingly, mortality was lower in women who conceived after breast cancer (39). For reasons of preserving health and fears of recurrence, these reassuring data can still be difficult for women to accept. About 50% of those who survive will want to try for pregnancy, but fewer than 10% will actually get pregnant – the detrimental effects of therapy and perhaps the time gap between diagnosis and the 'all clear' are probably very important here.

A 2018 review of the literature by Lambertini and colleagues examined this issue and listed several meta analyses which found reassuring results (40). A separate 7-year post-pregnancy follow-up study of 194 oestrogen receptor positive (ER+) and 139 oestrogen receptor negative (ER−) breast cancer patients who became pregnant compared them with 874 nonpregnant breast cancer controls of whom 292 had ER+ breast cancer (15). Disease-free survival (DFS) did not differ between the pregnant group and controls, nor did it differ between the pregnant ER+ and pregnant ER− groups. The length of time before getting pregnant was not a factor in length of DFS for any group. Those becoming pregnant less than 2 years after diagnosis did as well as those who waited longer. Live births were the outcome for 57% of all pregnancies, induced abortion was chosen in 27% of cases, and spontaneous abortion occurred in 6%. Results were unavailable for the remainder. These results are quite reassuring and suggest that pregnancy does not pose a major risk to breast cancer patients. Individual cases may warrant advice to delay pregnancy for more than 2 years, and some women may have to take ongoing medication, for example, tamoxifen, which is teratogenic, precluding pregnancy. The Lambertini study was unable to gather details on the use or otherwise of artificial reproduction methods.

The Carneiro review (40) noted a lack of IVF outcome evidence to draw upon in the literature, stating that studies were small and some were only case series. The results did suggest, however, that the outcome after embryo thawing and transfer was comparable to IVF in cases of infertility due to tubal disease or male aetiology. However, there are two studies which appear to be large enough to be of value. Luke et al. looked at over 53,000 women who had artificial reproductive technologies (ART) between 2004 and 2009 and identified 441 of these who had ART *after* cancer (41). One-third of the 441 were breast cancer survivors, so this study provides information on the pregnancy outcomes for other cancers. These were endocrine, melanoma and all female genital (cervix, uterus, ovary, vagina and vulva). The important finding was that cancer survivors who used their own eggs for ART were much less likely to have a live birth than noncancer patients (24.7% vs 47.7%). Cancer survivors who used donor eggs had almost the same live birth rates as noncancer patients (60.4% vs 64.5%). The outcomes were particularly poor for own egg use in breast cancer patients. Most breast cancer women used their own eggs, and in that group, the live birth rate was only 14.3%. It was not stated whether the patient's eggs had been frozen prior to cancer treatment or whether the eggs were collected post-treatment, but these figures are startling. Only eight breast cancer survivors used donor eggs, but the live birth rate here was equivalent to that for women without cancer (60% vs 65%, respectively) (42). The cancer type also exerted an effect. Those who had melanoma diagnoses were much more likely to have a live birth than those with breast cancer. Goldrat's study of pregnancy in breast cancer survivors was smaller (43). One hundred ninety-eight women became pregnant after treatment for breast cancer, 25 of whom became pregnant through ART; the rest were spontaneous conception. Full-term pregnancies occurred in similar proportions in both groups, from which data they concluded that pregnancy using ART is feasible in women with previous breast cancer.

Young women get other cancers as well as breast cancer; cervical cancer is the next most common. Cervical cancer presents additional challenges in that radiotherapy and even hysterectomy can be a part of the treatment. Fortunately, surgery alone (as in removing a portion of the cervix) can be sufficient for cervical cancers, with disease progression limited to and including Stage T2a2 (Tumour, Node, Metastasis [TNM] classification) or Stage IIA2 (Fédération Internationale de Gynécologie et d'Obstétrique [FIGO]) (without lymph node involvement), which implies clinically visible lesion more than 4.0 cm in greatest dimension (44). This is likely to leave a weakened and incompetent cervix requiring a cervical suture (and probably using an abdominal approach rather than vaginal) just to maintain the pregnancy, but fertility and the ability to conceive are unimpaired or only slightly reduced because of loss of cervical mucus (see Sections 3.3 and 4.7 for further information on cervical mucus). With more advanced stages of malignancy, radiotherapy and radical surgery are required. The usual practice is to remove the cervix, uterus and some of the pelvic wall (a radical hysterectomy), during which procedure the ovaries are moved and usually fixed to the inner aspect of the iliac crest, where they will be safe from radiotherapy. The ovarian blood supply can be damaged, either on that occasion, or more likely when they are later moved back into the pelvis. However, without their being moved back, they will be inaccessible to transvaginal oocyte recovery. Almost all those gynaecologists trained in the 1980s and 1990s in laparoscopic oocyte collection have stopped clinical work, making this option for oocyte recovery no longer feasible. Ideally, oocyte recovery would happen before radical clearance of the pelvis is undertaken and while the ovaries are in an undisturbed state, but the risk of spread of cancerous cells by attempting transvaginal oocyte collection makes this unsafe.

In women with lymphomas, it is not unusual to have spinal cord irradiation to treat all the lymph nodes in the area. The uterus will be included within the field of radiation because of its midline position. For these women, the blood supply and the ability of the endometrium to develop are likely to be impaired. The remaining function of the endometrium can be tested by giving oestrogen in different administration forms (locally we use both oral and patch treatment), seeking to stimulate endometrial growth.

There is no agreed way to stimulate endometrial tissue to prepare for implantation, something ably explained by Macken in a review paper detailing current knowledge on timing of and optimal endometrial preparation for embryo transfer (47). While there is variation in how long to give oestrogen for, ranging from 5 days to 2 weeks, and in my clinic 4 weeks, I am persuaded that given that the circumstances relate to postradiotherapy, 4 weeks is good practice to ensure that an adequate test of the endometrium is provided (and it is noteworthy that some approaches allow for up to 6 months or longer of therapy [48]). The concern from prolonged use of oestrogen is thromboembolism. One poor-quality retrospective study examined the role of vaginal sildenafil. When treated with sildenafil, more women (70%) had an endometrium thicker than 9 mm compared to those untreated (30%) (45). Implantation rates were markedly different in the two groups: 29% (treated) vs 2% (untreated) ($P < 0.01$). A pilot study compared the use of sildenafil to a control and found that blood flow improved and endometrial thickness increased (49). Data from the pentoxifylline and tocopherol studies (46) seem to suggest that all women responded to the treatment. For the plan laid out in Figure 9.3, it is worth noting that the vaginal sildenafil had to be manufactured specially and is not available commercially at present. In such a plan for endometrial development, some clinicians use aspirin because of its effect on blood vessels. However, a review of the literature on aspirin and the endometrium provides no evidence to support that use (50).

Programmed frozen embryo transfer protocol: endometrial stimulation following radiotherapy

Setting	Division of Women and Children's Services / Reproductive Medicine / IUI clinic
For Staff	Staff within the Reproductive Medicine Department
Patients	Patients who had had pelvic radiotherapy and similar treatments

Women exposed to therapeutic doses of pelvic radiotherapy in the course of treatment for cancer will have had significant damage to the endometrium because of this.

Treatment will last at least 4 weeks and up to 8 weeks, using the regimen below.

Plan of treatment:

1. Oestrogen stimulation to start
   i. Oestradiol valerate (Progynova), 2 mg four times a day for 4 weeks
   ii. Oestradiol patches 100 microgram twice weekly (Monday and Thursday) for 4 weeks
2. Review on transvaginal scan at 4 weeks
3. If endometrium growing >5 mm continue the above but add in Cyclogest for 2 weeks and then withdraw
4. Wait for withdrawal bleed
5. In the absence of growth of 5 mm or more, add in aspirin (75–150 mg by BMI) and consider addition of sildenafil or similar
6. Review on transvaginal scan at 4 weeks
7. If growing, then revert to point 3 for further management. If not, discontinue treatment

Drugs for this regimen:

Oestradiol valerate (Progynova), 2 mg per tablet:	114 tablets
Oestradiol patches 100 mg (FemSeven/Progynova TS 100)	8 patches
Progesterone (Cyclogest), 400 mg vaginal pessary	58 pessaries
Sildenafil 25 mg pessaries, 4 times a day	112 pessaries
Aspirin 75–150 mg	28 tabs (OTC)

Effect of vaginal sildenafil on the outcome of *in vitro* fertilization (IVF) after multiple IVF failures attributed to poor endometrial development. Fertil Steril. 2002 Nov;78(5):1073-6.

Figure 9.3 Stimulation of the endometrium following radiotherapy to evaluate if the endometrium will respond sufficiently to facilitate embryo or blastocyst transfer.

9.4 SOCIAL INDICATIONS FOR FERTILITY PRESERVATION: WHAT HAPPENS TO FERTILITY WITH INCREASING AGE?

In the developed world, as women have an increasing role in the workplace, they face a decision about when to have their children. Addressed in Chapter 8, this stems from the increasing costs of living, of accommodation and of child care, and the consequent need to develop well-paying and secure career positions while young, In the United Kingdom, the average age at first pregnancy is 30 (51); in the United States, the average age is lower (26.6 years), but has been increasing yearly since 2002, when the average was 24.9 years (52). High-profile technology companies (Facebook and Google among some) have been advertising opportunities for their female employees to store their oocytes for free or at low cost to enable employees to preserve their fertility. These companies particularly value the imagination and innovative ability of youth, reflected in their average working age of their employees being much lower than in many other work environments.

9.4.1 Natural changes in fertility potential

Some of the data referred to in this chapter have been discussed earlier in this book – in Chapter 2 (see Sections 2.1.2 and 2.4) and Chapter 8 (see Section 8.4.2). Fertility for both women and men declines with time. For women, the decline is steady and slow, year by year, until women reach the age of 37–38 years. After that, the decline in fertility accelerates. Evidence for this is available from scientific models assessing ovarian follicle numbers and from population data looking at birth rates. A frequently cited study by Guttmacher, examined 209 women from the Hutterite community in North America, which does not use contraception. In that population, he demonstrated the naturally occurring age related reduction in fertility (53). Michael Faddy more recently conducted work on ovarian follicles (54). As already mentioned in Chapter 3, he showed that follicles were lost at a specific rate until the overall numbers fell to a critical figure of 25,000, which happens at the age of 37.5 years. Thereafter, there was a much faster rate of follicle loss, until numbers fell to approximately 1000 somewhere around 51 years, coincident with menopause. Separately, work has been done on the quality of oocytes as women get older, showing a reduction in quality largely related to degenerative oocytes (55).

For men, there is a decline in fertility caused by a fall in testosterone production in the testis, causing a gradual reduction in the number of sperm. This usually does not become evident until after the age of 55 (56,57). In a large retrospective study of women aged 25–45, Rochebrochard and Thonneau found that for women aged 35–40 years, the likelihood of getting pregnant was less if the male partner was over 40 (58). Interestingly, however, even if the male partner was over 40, the risk of infertility was not increased significantly when the woman was aged <34 years. These findings were replicated in other studies. The European Fecundability Study (59) supports the view that men aged 40 and over have a lower chance of conception than men aged 35 or under with the same frequency of intercourse, and data from IVF cycles showed lower pregnancy rates following both IVF and ICSI associated with paternal age in excess of 51 (60).

In Chapter 7, the issue was addressed under lifestyle choices, recognising that more women are 'choosing' to delay childbearing until their thirties, and sometimes later. The reasons are complex. The common assumption, challenged later in the chapter is that women are selfishly or unknowingly jeopardising their chances of bearing children to pursue career, money and freedom to experience life unencumbered, while also being too particular to form a lasting supportive partnership in which to rear children (61). The falling national population figures and the rising tax burden on fewer and fewer workers is attributed to the younger generation not taking on their responsibility to society as a whole (62). To correct this, there is a push, often by public health or medical bodies, to help women realise the consequences of their 'ticking clock'. Many public health campaigns have tried to address this without recognising how multifactorial the problem is. One stark visual impact is presented as Figure 9.4 – 'Break before it's too late' (break the cycle of declining fertility and consider doing something active to preserve your fertility). Telling women in their 20s and 30s

Figure 9.4 Break before it's too late: A theoretical campaign to remind young women of their declining fertility. (By courtesy of Elizabeth Ewing.)

is maybe too late – it is likely that this needs to be introduced into sex education in late second-level education, balancing the messages of 'avoid pregnancy' with '… but not for too long'.

Indeed, many surveys do show that a surprising proportion of women (and men) are ignorant of the limited fertility span for women, or if they are aware of it, they have unrealistic expectations of the success rates of artificial reproduction techniques. For example, an Australian population survey of over 350 childless couples found that most had little understanding of the impact of their age of their fertility, and would have planned differently for their life had they known so (63). Reproductive medicine specialists are also only too familiar with the distress felt by their patients when they realise that the opportunity to conceive has passed them by, and some of the medical impetus to 'educate' young women about the time limit of their fertility derives from these sad clinical encounters.

People live and make choices within a rubric of complex and interconnected influences. Sociological discussion is outside the scope of this text, but a good account of the social, economic and ideological factors which shape the choices women and men make about having children can be found in a paper commissioned by the European Society for Human Reproduction and Embryology, "Why Do People Postpone Parenthood? Reasons and Social Policy Incentives" (64). The paper touches on many issues. Unemployment is high in many countries, and even where the statistics say unemployment is low, jobs are low paid and insecure. In some countries, having children is a signal to employers to let their female worker go, even if employment law theoretically protects them. Childcare costs, as well as the cost of living, are very high. Women who obtain a primary degree in their 20s are often encouraged by their mentors, both academic and workplace, to pursue further education to maximise their employability potential in the future and to increase income. Gender equality is still a societal and political issue, and women are encouraged to advance to the top of the career ladder and break through to the higher levels of industry and academia to redress the unequal representation of women in positions of power. Having children is postponed until the career goal is achieved, because women understand the demands motherhood will place on them, and the difficulties meeting the child's needs in the current environment. The ESHRE paper notes that it is the responsibility of government policy to change the working environment for both sexes. The goal is to make it possible for both women and men to incorporate child rearing into their working lives and enable affordable and good child care to accommodate working women, or those in continuing education and training, so that they can have children earlier, in their mid-20s.

These are the policy, public health and public education debates. Where is society heading while these debates play out? There already is public awareness that women may have problems conceiving or carrying a pregnancy if they wait too long, with a lot of coverage of the topic in different media outlets (65).

There is the much-discussed offer by Apple and Facebook to include egg freezing in the benefits package available to their female employees who are pursuing a career path with them. More and more people in their late 20s and 30s do know the 'facts of fertility' and have tentative plans about how to ensure they have children when the time is right. An international study of university staff and undergraduate students (childless women aged 28–35 years in the United Kingdom, United States and Canada – 23 universities) found that many were considering using fertility preservation to achieve parenthood, while they were also expecting to have children at a later age (66). Irish women were asked the enigmatic question, 'What do women want *in the area of fertility preservation?*' More than 75% of the respondents thought a test of ovarian reserve was important information to access, while also voicing the idea that oocyte preservation was a woman's right in order to preserve her fertility (62% of respondents) (67). A survey from the United States using a social media platform reached a cohort of 1000 women aged 21–45 years (68). More than 80% said they were aware of oocyte freezing for fertility preservation, although most didn't understand what that entailed; 30% said they would be prepared to do it and to pay for it.

With the increasing availability of technologies to successfully freeze eggs, the opportunity to, as it were, insure against the future, is now open to women and it is being marketed to them. There seems little doubt that 'social' fertility preservation will be – already is – a health choice for many.

The options that women have for fertility preservation have already been described: oocyte freezing, embryo freezing and ovarian tissue freezing, and the procedures and success rates are detailed in Section 9.2.3. The commercial possibilities of this new market and the money to be made from providing these services have encouraged the emergence of new clinics, all promoting the advantages of this kind of 'insurance plan', and of 'taking charge of your family's future'. A 2014 Huffington Post article by the medical director of a fertility centre is an example of information/ advertisement on egg cryopreservation in the United States (69).

The reaction within the medical profession has been twofold. There is still a hope that by informing young women of the realities of time-limited fertility and exhorting them to make different lifestyle choices in their own interests, behaviours will change. On the other hand, there is concern about the ethics and morality of marketing fertility preservation to people who are not ill and not suffering from a medical condition, selling medical interventions to those who are not ill. There is the concern that marketing will not be frank about the success rates of live births, about the risks to the individual undergoing oocyte collection and embryo transfer or about the complications attending pregnancy for women of mature years (70). Particularly in Europe, there is a distaste for and distrust of commercialised medicine amongst the medical profession, paternalistically fearing that the motivation of profit will exploit and take advantage of a lack of knowledge and vulnerability. The market for 'social' freezing theoretically is limited only by the high cost of the procedures. The ethical aspects of speculative freezing of eggs or embryos in case they are needed in the future; the unseemly prospects of the individual failing to meet payments for storage, resulting in destruction of the tissue, the growth of a market for unwanted eggs which can be sold on; more and more disputes about ownership of the gametes or embryos if a couple splits up or the individual dies – all these scenarios generate a lot of discussion and often dismay.

Men, who do not need to worry about declining fertility until into their late 40s, tend not to seek to preserve their own sperm, unless they know that they have a specific problem with their sperm. Very occasionally, men who are about to have a vasectomy will enquire and may proceed with sperm preservation to provide an insurance against potential changes of view in later years.

9.4.1.1 Clinical response to declining fertility and demand for fertility preservation

So as clinicians, what can we do? I suggest we should use every opportunity given to us to

- Educate young women and men about the true nature of their age-related fertility.
- When we encounter women in a clinical setting who may raise this tangentially, we should have accurate and up-to-date information on how to access local oocyte preservation services.
- Don't discourage women from exploring preservation – provide them with data.
- Ensure that our trainees and medical and nursing colleagues are aware of this information.

National bodies with expertise in this area (for instance, the American Society of Reproductive Medicine) have accepted that these treatments are a suitable way for women to protect their fertility, although they express caution about the efficacy and long-term effects of oocyte preservation, something women should be apprised of before starting treatment (71).

9.5 SUMMARY

This chapter explores both medical and social freezing of oocytes. Social freezing is a complex issue because of the pressure we, our society and our work environment, place on younger people to balance careers and parenthood. Methods of preservation of oocytes with some information on its biology, in whatever scenario, are discussed, together with expectations for success.

REFERENCES

1. Stachs A, Hartmann S, Gerber B. Preservation of fertility or ovarian function in patients with breast cancer or gynecologic and internal malignancies. *Geburtshilfe und Frauenheilkunde*. 2017;77(8):861–9.
2. Thomson AB, Critchley HO, Wallace WH. Fertility and progeny. *European Journal of Cancer*. 2002;38(12):1634–44; discussion 1645–6.
3. Relander, T, Cavallin-Stahl E, Garwicz ES, Olsson AM, Willen M. Gonadal and sexual function in men treated for childhood cancer. *Medical and Pediatric Oncology*. 2000;35(1):52–63.
4. Wallace WH, Shalet SM, Lendon M, Morris-Jones PH. Male fertility in long-term survivors of childhood acute lymphoblastic leukaemia. *Intetnation Journal of Andrology*. 1991;14(5):312–9.
5. Birkebaek NH, Fisker S, Clausen N, Tuovinen V, Sindet-Pedersen S, Christiansen JS. Growth and endocrinological disorders up to 21 years after treatment for acute lymphoblastic leukemia in childhood. *Medical and Pediatric Oncology*. 1998;30(6):351–6.
6. Liesner RJ, Leiper AD, Hann IM, Chessells JM. Late effects of intensive treatment for acute myeloid leukemia and myelodysplasia in childhood. *Journal of Clinical Oncology*. 1994;12(5):916–24.
7. Teinturier C, Hartmann O, Valteau-Couanet D, Benhamou E, Bougneres PF. Ovarian function after autologous bone marrow transplantation in childhood: High-dose busulfan is a major cause of ovarian failure. *Bone Marrow Transplantion*. 1998;22(10):989–94.
8. Papadakis V, Vlachopapadopoulou E, Van Syckle K, Ganshaw L, Kalmanti M, Tan C, Sklar C. Gonadal function in young patients successfully treated for Hodgkin disease. *Medical and Pediatric Oncology*. 1999;32(5):366–72.
9. Wallace WH, Shalet SM, Hendry JH, Morris-Jones PH, Gattamaneni HR. Ovarian failure following abdominal irradiation in childhood: The radiosensitivity of the human oocyte. *British Journal of Radiology*. 1989;62(743):995–8.
10. Meirow D, Biederman H, Anderson RA, Wallace WH. Toxicity of chemotherapy and radiation on female reproduction. *Clinical Obstetrics and Gynecology*. 2010;53(4):727–39.
11. Wallace WH, Shalet SM, Hendry JH, Morris-Jones PH, Gattamaneni HR. Ovarian failure following abdominal irradiation in childhood: The radiosensitivity of the human oocyte. *British Journal of Radiology*. 1989;62(743):995–8.
12. Wallace WH, Thomson AB, Kelsey TW. The radiosensitivity of the human oocyte. *Human Reproduction (Oxford, England)*. 2003;18(1):117–21.
13. Kim J, Kim KH, Mersereau JE. Building a successful fertility preservation program at a major cancer center. *Journal of Gynecologic Oncology*. 2014;25(2):148–54.
14. Herrero HB, Chan PTK. Children after cancer? Meeting male patients' fertility needs during cancer care. *EMJ Reproductive Health*. 2018;4(1):76–82.
15. Lambertini M, Kroman N, Ameye L, Cordoba O, Pinto A, Benedetti G et al. Long-term safety of pregnancy following breast cancer according to estrogen receptor status. *Journal of the National Cancer Institute*. 2018;110(4):426–9.

16. Duffy C, Allen SM, Dube C, Dickersin K. Oncologists' confidence in knowledge of fertility issues for young women with cancer. *Journal of Cancer Education: Official Journal of the American Association for Cancer Education*. 2012;27(2):369–76.
17. Ortin TT, Shostak CA, Donaldson SS. Gonadal status and reproductive function following treatment for Hodgkin's disease in childhood: The Stanford experience. *International Journal of Radiation Oncology, Biology, Physics*. 1990;19(4):873–80.
18. Blumenfeld Z. How to preserve fertility in young women exposed to chemotherapy? The role of GnRH agonist cotreatment in addition to cryopreservation of embrya, oocytes, or ovaries. *Oncologist*. 2007;12(9):1044–54.
19. Munhoz RR, Pereira AA, Sasse AD, Hoff PM, Traina TA, Hudis CA et al. Gonadotropin-releasing hormone agonists for ovarian function preservation in premenopausal women undergoing chemotherapy for early-stage breast cancer: A systematic review and meta-analysis. *JAMA Oncology*. 2016;2(1):65–73.
20. Oktay K, Briggs D, Gosden RG. Ontogeny of follicle-stimulating hormone receptor gene expression in isolated human ovarian follicles. *Journal of Clinical Endocrinology and Metabolism*. 1997;82(11):3748–51.
21. Moore H, Unger J, Phillips K. Phase III trial (Prevention of Early Menopause Study [POEMS]-SWOG S0230) of LHRH analog during chemotherapy (CT) to reduce ovarian failure in early-stage, hormone receptor-negative breast cancer: An international intergroup trial of SWOG, IBCSG, ECOG, and CALGB. *Journal of Clinical Oncology: Official Journal of the American Society of Clinical Oncology*. 2014;32(5s (suppl; abstr LBA505)).
22. Gerber B, Ortmann O. Prevention of Early Menopause Study (POEMS): Is it possible to preserve ovarian function by gonadotropin releasing hormone analogs (GnRHa)? *Archives of Gynecology and Obstetrics*. 2014;290(6):1051–3.
23. Moore HCF, Unger JM, Phillips K-A, Boyle F, Hitre E, Moseley A et al. Final analysis of the Prevention of Early Menopause Study (POEMS)/SWOG intergroup S0230. *JNCI: Journal of the National Cancer Institute*. 2019:111:210–3.
24. Goldman RH, Racowsky C, Farland LV, Munne S, Ribustello L, Fox JH. Predicting the likelihood of live birth for elective oocyte cryopreservation: A counseling tool for physicians and patients. *Human Reproduction (Oxford, England)*. 2017;32(4):853–9.
25. Practice Committee of American Society for Reproductive Medicine. Fertility preservation in patients undergoing gonadotoxic therapy or gonadectomy: A committee opinion. *Fertility and Sterility*. 2013;100(5):1214–23.
26. Shapira M, Raanani H, Meirow D. IVF for fertility preservation in breast cancer patients–efficacy and safety issues. *Journal of Assisted Reproduction and Genetics*. 2015;32(8):1171–8.
27. Shi Q, Xie Y, Wang Y, Li S. Vitrification versus slow freezing for human ovarian tissue cryopreservation: A systematic review and meta-analysis. *Scientific Reports*. 2017;7(1):8538.
28. Dalman A, Deheshkar Gooneh Farahani NS, Totonchi M, Pirjani R, Ebrahimi B, Rezazadeh Valojerdi M. Slow freezing versus vitrification technique for human ovarian tissue cryopreservation: An evaluation of histological changes, WNT signaling pathway and apoptotic genes expression. *Cryobiology*. 2017;79:29–36.
29. Kokotsaki M, Mairhofer M, Schneeberger C, Marschalek J, Pietrowski D. Impact of vitrification on granulosa cell survival and gene expression. *Cryobiology*. 2018;85:73–8.
30. Donnez J, Dolmans MM, Pellicer A, Diaz-Garcia C, Sanchez Serrano M, Schmidt KT et al. Restoration of ovarian activity and pregnancy after transplantation of cryopreserved ovarian tissue: A review of 60 cases of reimplantation. *Fertility and Sterility*. 2013;99(6):1503–13.
31. Wallace WH, Anderson RA, Irvine DS. Fertility preservation for young patients with cancer: Who is at risk and what can be offered? *The Lancet Oncology*. 2005;6(4):209–18.
32. American Cancer Society. Cancers that Develop in Children 2016 [cited 2019 Jan 26]. Available from: https://www.cancer.org/cancer/cancer-in-children/types-of-childhood-cancers.html.

33. Trottmann M, Becker AJ, Stadler T, Straub J, Soljanik I, Schlenker B et al. Semen quality in men with malignant diseases before and after therapy and the role of cryopreservation. *European Urology*. 2007;52(2):355–67.
34. Colpi GM, Contalbi GF, Nerva F, Sagone P, Piediferro G. Testicular function following chemo-radiotherapy. *European Journal of Obstetrics, Gynecology, and Reproductive Biology*. 2004;113(Suppl 1):S2–6.
35. van Casteren NJ, Dohle GR, Romijn JC, de Muinck Keizer-Schrama SM, Weber RF, van den Heuvel-Eibrink MM. Semen cryopreservation in pubertal boys before gonadotoxic treatment and the role of endocrinologic evaluation in predicting sperm yield. *Fertility and Sterility*. 2008;90(4):1119–25.
36. Onofre J, Baert Y, Faes K, Goossens E. Cryopreservation of testicular tissue or testicular cell suspensions: A pivotal step in fertility preservation. *Human Reproduction Update*. 2016;22(6):744–61.
37. Gies I, De Schepper J, Tournaye H. Progress and prospects for fertility preservation in prepubertal boys with cancer. *Current Opinion in Endocrinology, Diabetes, and Obesity*. 2015;22(3):203–8.
38. Jonat W, Kaufmann M, Sauerbrei W, Blamey R, Cuzick J, Namer M et al. Goserelin versus cyclophosphamide, methotrexate, and fluorouracil as adjuvant therapy in premenopausal patients with node-positive breast cancer: The Zoladex Early Breast Cancer Research Association Study. *Journal of Clinical Oncology: Official Journal of the American Society of Clinical Oncology*. 2002;20(24):4628–35.
39. Azim HA, Jr., Santoro L, Pavlidis N, Gelber S, Kroman N, Azim H et al. Safety of pregnancy following breast cancer diagnosis: A meta-analysis of 14 studies. *European Journal of Cancer (Oxford, England: 1990)*. 2011;47(1):74–83.
40. Carneiro MM, Cota AM, Amaral MC, Pedrosa ML, Martins BO, Furtado MH et al. Motherhood after breast cancer: Can we balance fertility preservation and cancer treatment? A narrative review of the literature. *JBRA Assisted Reproduction*. 2018;22(3):244–52.
41. Luke B, Brown MB, Spector LG, Missmer SA, Leach RE, Williams M et al. Cancer in women after assisted reproductive technology. *Fertility and Sterility*. 2015;104(5):1218–26.
42. Luke B, Brown MB, Missmer SA, Spector LG, Leach RE, Williams M et al. Assisted reproductive technology use and outcomes among women with a history of cancer. *Human Reproduction (Oxford, England)*. 2016;31(1):183–9.
43. Goldrat O, Kroman N, Peccatori FA, Cordoba O, Pistilli B, Lidegaard O et al. Pregnancy following breast cancer using assisted reproduction and its effect on long-term outcome. *European Journal of Cancer (Oxford, England: 1990)*. 2015;51(12):1490–6.
44. Cibula D, Potter R, Planchamp F, Avall-Lundqvist E, Fischerova D, Haie Meder C et al. The European Society of Gynaecological Oncology/European Society for Radiotherapy and Oncology/European Society of Pathology guidelines for the management of patients with cervical cancer. *Radiotherapy and Oncology: Journal of the European Society for Therapeutic Radiology and Oncology*. 2018;127(3):404–16.
45. Sher G, Fisch JD. Effect of vaginal sildenafil on the outcome of *in vitro* fertilization (IVF) after multiple IVF failures attributed to poor endometrial development. *Fertility and Sterility*. 2002;78(5):1073–6.
46. Teh WT, Stern C, Chander S, Hickey M. The impact of uterine radiation on subsequent fertility and pregnancy outcomes. *BioMed Research International*. 2014;2014:8.
47. Mackens S, Santos-Ribeiro S, van de Vijver A, Racca A, Van Landuyt L, Tournaye H et al. Frozen embryo transfer: A review on the optimal endometrial preparation and timing. *Human Reproduction*. 2017;32(11):2234–42.
48. Letur-Konirsch H, Delanian S. Successful pregnancies after combined pentoxifylline-tocopherol treatment in women with premature ovarian failure who are resistant to hormone replacement therapy. *Fertility and Sterility*. 2003;79(2):439–41.

49. Takasaki A, Tamura H, Miwa I, Taketani T, Shimamura K, Sugino N. Endometrial growth and uterine blood flow: A pilot study for improving endometrial thickness in the patients with a thin endometrium. *Fertility and Sterility*. 2010;93(6):1851–8.
50. Lebovitz O, Orvieto R. Treating patients with 'thin' endometrium - An ongoing challenge. *Gynecological Endocrinology: Official Journal of the International Society of Gynecological Endocrinology*. 2014;30(6):409–14.
51. Office for National Statistics. *Births by Parents' Characteristics in England and Wales: 2015*. London: Office for National Statistics; 2015. Available from: https://www.ons.gov.uk/peoplepopulationandcommunity/birthsdeathsandmarriages/livebirths/bulletins/birthsbyparentscharacteristicsinenglandandwales/2015.
52. Martin J, Hamilton B, Osterman M, Driscoll A, Drake P. *Births: Final Data for 2016*. Hyattsville, MD: National Center for Health Statistics; 2018.
53. Guttmacher AF. Factors affecting normal expectancy of conception. *Journal of the American Medical Association*. 1956;161(9):855–60.
54. Faddy MJ, Gosden RG, Gougeon A, Richardson SJ, Nelson JF. Accelerated disappearance of ovarian follicles in mid-life: Implications for forecasting menopause. *Human Reproduction (Oxford, England)*. 1992;7(10):1342–6.
55. Lim AS, Tsakok MF. Age-related decline in fertility: A link to degenerative oocytes? *Fertility and Sterility*. 1997;68(2):265–71.
56. Almeida S, Rato L, Sousa M, Alves MG, Oliveira PF. Fertility and sperm quality in the aging male. *Current Pharmaceutical Design*. 2017.
57. Sigman M. Introduction: What to do with older prospective fathers: The risks of advanced paternal age. *Fertility and Sterility*. 2017;107(2):299–300.
58. Rochebrochard EDL, Thonneau P. Paternal age ≥40 years: An important risk factor for infertility. *American Journal of Obstetrics and Gynecology*. 2003;189(4):901–5.
59. Dunson DB, Baird DD, Colombo B. Increased infertility with age in men and women. *Obstetrics and Gynecology*. 2004;103(1):51–6.
60. Chapuis A, Gala A, Ferrieres-Hoa A, Mullet T, Bringer-Deutsch S, Vintejoux E et al. Sperm quality and paternal age: Effect on blastocyst formation and pregnancy rates. *Basic and Clinical Andrology*. 2017;27:2.
61. Renzetti E. Why Childless People Are Persecuted? [Newspaper]. Toronto, ON, Canada: *The Globe and Mail*; 2013. Available from: https://www.theglobeandmail.com/life/parenting/why-childless-people-are-persecuted/article12005541/.
62. Kotkin J, Siegel H. Where have all the babies gone? *Newsweek*. 2013.
63. Evans A, de Lacey S, Tremellen K. Australians' understanding of the decline in fertility with increasing age and attitudes towards ovarian reserve screening. *Australian Journal of Primary Health*. 2018.
64. Mills M, Rindfuss RR, McDonald P, te Velde E. Why do people postpone parenthood? Reasons and social policy incentives. *Human Reproduction Update*. 2011;17(6):848–60.
65. Sinclair C. *Fertility Debate: How Late Is Too Late*? London: Vogue; May 2012.
66. Ter Keurst A, Boivin J, Gameiro S. Women's intentions to use fertility preservation to prevent age-related fertility decline. *Reproductive Biomedicine Online*. 2016;32(1):121–31.
67. O'Brien Y, Martyn F, Glover LE, Wingfield MB. What women want? A scoping survey on women's knowledge, attitudes and behaviours towards ovarian reserve testing and egg freezing. *European Journal of Obstetrics and Gynecology and Reproductive Biology*. 2017;217:71–6.
68. Milman LW, Senapati S, Sammel MD, Cameron KD, Gracia C. Assessing reproductive choices of women and the likelihood of oocyte cryopreservation in the era of elective oocyte freezing. *Fertility and Sterility*. 2017;107(5):1214–22.e3.
69. Beltsos A. 10 Realities of Freezing Your Eggs: What Women Need to Know. *Huffington Post*; 2014. Available from: https://www.huffingtonpost.com/dr-angeline-beltsos/10-realities-of-freezing-_b_5656344.html.

70. Petropanagos A, Cattapan A, Baylis F, Leader A. Social egg freezing: Risk, benefits and other considerations. *CMAJ: Canadian Medical Association Journal = Journal de L'Association Medicale Canadienne.* 2015;187(9):666–9.
71. Practice Committee of American Society for Reproductive Medicine. Planned oocyte cryopreservation for women seeking to preserve future reproductive potential: An Ethics Committee opinion. *Fertility and Sterility.* 2018;110(6):1022–8.

Global perspectives on reproductive medicine

10

OVERVIEW

This chapter deals with infertility from a global aspect, providing information on how infertility presents differently all over the world. One section focuses on the attitudes of major world religions to infertility and to its treatment, seeking to enlighten readers on how faith can affect attitudes of patients, which we cannot ignore. Finally, the author highlights how little progress the developed world has made in giving the promised international support to developing counties to offer interventions to treat infertility in all its presentations in those countries.

Almost 200 countries are recognised by the United Nations as official entities. Countries are divided into developed and developing, based on their economy and whether it is pre- or postindustrial and whether the country is dependent on agriculture but not industry. Standards of provision and availability of public and health services cannot be inferred from the label 'developed country'. Consideration of the differences between the United States and Europe show this, although the difference between them is a result of complex political and sociological ideologies. The provision of infertility/reproductive medicine services across these developed counties is just as, perhaps even more, uneven than the usual health services. Even within countries, there can be marked discrepancies. The United Kingdom is one example, where there are major differences in access to and provision of infertility/reproductive medicine services depending on where the patient lives, termed the postcode lottery by UK citizens. In the United Kingdom, lifestyle factors are used by different health authorities in the country as qualifying or excluding criteria for allowing fertility treatment. However, no matter how unevenly distributed things are in developed countries, access to infertility/reproductive medicine services in less well-developed countries is extremely variable, with a wealth of data to show that assisted conception clinics are fewer in those countries (1,2).

10.1 WHAT DO WE KNOW ABOUT INFERTILITY ACROSS THE GLOBE?

10.1.1 Prevalence

One large study of prevalence looked at 277 surveys from 1990 to 2010 (4). From this analysis, they showed that globally, among women aged 20–44 who were exposed to the possibility of pregnancy:

- 1.9% were unable to attain a live birth (primary infertility)
- 10.5% were unable to have a second or further child (secondary infertility)

Trying to find accurate global data on the numbers of couples is frustrating, made worse by 'gaps in survey data for some countries and the use of proxies to determine exposure to pregnancy' (4). Overall, as population numbers increase, the absolute numbers of infertile people will increase – though it is interesting to note there is 'reduced child-seeking behaviour result(ing) in a reduction of primary infertility among all women from 1.6% to 1.5% and a reduction of secondary infertility among all women' (4).

This balance of primary and secondary infertility is noteworthy. In the developed world, primary infertility prevalence generally exceeds that of secondary infertility (5). This is attributed to the likelihood that most of the major (and insurmountable) barriers to fertility will be met as primary infertility and those who do get pregnant will be the ones with less difficult reasons for infertility and consequently more likely to conceive again – many people with very severe problems of sperm disorder or blocked tubes will never overcome those, while those who have conceived once will generally find it easier to conceive a second time. If rates of secondary infertility are higher, it suggests that something is happening after the first pregnancy to reduce fertility. Lunenfeld and van Steirteghem suggest that the biggest factors leading to the high secondary infertility rates in these countries (sub-Saharan Africa, Cambodia, India, Indonesia) are sexually transmitted infections and medical interventions under unhygienic conditions, particularly postpartum. One of the biggest contributors in the STI group is HIV infection. There appears to be good reason for this view, as rising trends in HIV and other STIs coincide with increasing rates in secondary infertility (once again, associative not causative) (6). Reduced fecundity in individuals infected with HIV is well documented. Several factors may contribute to this: reduced coital frequency (due to ill health) and increased contraceptive use, including barrier methods to minimise viral transfer.

10.1.2 Provision of reproductive/fertility services

In many countries of the world, but particularly in less developed countries, the effects of infertility are negative for women, who experience humiliation and stigma because of it more than their male partners. Infertility is still seen as being a problem of the woman's 'making', and she can be shunned by the family, mistreated or divorced (8). In many societies, she may be ostracised and be at risk of starving (3). Multiple authors point to the situation whereby many surveys and investigations have all led to the conclusion that, certainly in the developing world, access to infertility and assisted conception services are minimal to nonexistent, and often when available, limited by the substantial costs to only those who can afford to pay (3,7,8).

As one of its Millennium Development Goals, the UN agreed in 1994 to provide 'universal sexual and reproductive' health care by the year 2015. This goal included family planning and other reproductive health services, such as emergency obstetric care, and diagnosis and treatment of sexually transmitted infections and diseases (7). They did not include provision of 'infertility health care' or reproductive medicine in the sense it is referred to through this book. Perhaps in response to that lack, in 2001 the World Health Organization proposed that infertility be considered a global health problem and called for efforts to adapt assisted reproduction techniques so as to make them accessible in low resource countries (27). In his paper on this, Ombelet (who has been one of the forces behind this drive) asks (the developed world) 'Do we care?' (1). In his paper, he goes on to highlight the two main reasons for not prioritizing infertility health care - overpopulation and resources, and expands further on these themes.

Good-quality data on infertility is lacking in many developing countries, and international non-government organisations (NGOs) working in those countries put more emphasis on contraception to reduce overall birth rates, rather than fertility treatments. The need to provide infertility services in developing countries is much debated, between of the expected overpopulation of developing countries and the drive to reduce the high birth rate using contraception, as opposed to the uncertainty of the effect on population numbers if low-cost infertility treatment (including assisted conception) were available.

The introduction of an infertility/fertility service or programme would need support from local government and some international support. In 2001, the World Health Organization called for the integration of infertility services into sexual and reproductive health programmes in all countries. This has not progressed much due to a combination of barriers. These are, in no particular order, budgetary constraints; the loss of medical, nursing and scientific staff to more lucrative posts elsewhere and lack of political commitment internationally (3).

An encouraging, and even inspiring, project is the Walking Egg Project, a not-for-profit organisation supported by the European Society of Human Reproduction and Embryology (9). This initiative seeks to provide low-cost IVF (approximately €200) with the minimum of technological support and intervention. Pilot projects have been developed in Gent, Belgium, with some success, and the first baby was born in Accra, Ghana in August, 2017 (26).

10.2 ATTITUDES OF THE MAJOR WORLD RELIGIONS TOWARDS INFERTILITY AND ASSISTED CONCEPTION

Your patients will appreciate it and it will assist their effective management if you are aware of the possible impact their faith, culture or personal beliefs have upon their experience of infertility and the acceptability of different treatments. The doctor's knowledge of world religions and cultures need not be encyclopaedic, but some understanding will pay dividends when management plans are being discussed. Individuals who don't adhere to any particular faith may have their own personal views on the morality or acceptability of different treatment options, and they also will appreciate your interest. Asking about cultural or faith beliefs may not be second nature to doctors, but clinicians will benefit in the long run from doing so. Questions about how they feel about any of the proposed procedures is important. So is the follow-up question – What will your parents/family think about what you're doing, and does it concern you? The follow-up question, even more than the opening question, can open up the family and cultural attitudes to the issue, which the couple themselves may not hold, but which will nevertheless have an impact on them as they move through treatment. Asking questions, being open and interested and being alert for points of tension or unease is key to a successful engagement.

The following religions represent the faith groups most frequently found in the United Kingdom, where I work: Christianity (Orthodox, Catholic and Protestant/Evangelical), Hinduism, Islam (Sunni and Shi'a) and Judaism.

10.2.1 Hinduism

In Hinduism, beliefs are based on the teachings of the sacred Vedas – a set of ancient texts about 3000 years old. Hinduism also incorporates a range of philosophies – polytheism, atheism, pantheism and monotheism. Karma, the fate dealt to you as a result of previous incarnations, and dharma, your personal duty, are also important principles. Regarding fertility, much emphasis is placed on having children, but particularly on having a male offspring. Dharma incorporates moral and ethical duty. In classical Hinduism, the human embryo is accorded special status and protection. In addition, there is no distinction made between conception and the embryo which develops and the more recognizable human shape seen in later pregnancy (10). Children are incarnations of past lives, and therefore their absence prevents that reincarnation from being carried through.

In the Hindu tradition, assisted conception is viewed as an act which is harmonious between the sacred and the secular. In this view, the limitations of the science of fertility are accepted, meaning that while failure to conceive will bring personal disappointment to a couple, within the culture there is an understanding that infertility is a normal part of life.

There is little in Hindu thought that disagrees with most practices in assisted conception. The classical Hindu position has over the centuries been influenced from outside, and may be considered to be more flexible as a result. Positions on details of assisted reproduction are flexible and open to debate, allowing for a greater breadth of acceptance of any technique within Hinduism. Ideally, the oocytes and sperm used in treatment should be from the married couple themselves. Sperm donation is acceptable, particularly if the donor is a close relative to the infertile man in the relationship (12). In summary, a couple from the Hindu tradition seeking fertility assistance will not usually be in conflict with any set restrictions on assisted conception by their culture. While fertility is highly regarded, there is an understanding that some aspects of life are not under the control of science and the individual.

10.2.2 Islam

In Muslim societies, family and fertility are esteemed highly. The Quran, Islam's sacred text, has various citations which show the importance of many children in the life of the Muslim community. Not being able to get children with a wife is an acceptable reason for separation and for a man to take a second wife. Couples may already have three or four children, but attend the infertility clinic because of difficulties in having more.

In 1980 (2 years after the delivery of Louise Brown), the Grand Imam of Al Azhar (he is the head of Al Azhar, a Sunni religious university in Cairo) issued a proclamation enshrining several principles for the use of assisted conception with Islam – which are still relevant and are paraphrased here (13):

- Since marriage is a contract between the wife and husband during the span of their marriage, no third party intrudes into the marital functions of sex and procreation.
- A third party is not acceptable, whether he or she is providing a sperm, an egg, an embryo or a uterus.
- If the marriage contract has come to an end because of divorce or death of the husband, artificial reproduction cannot be performed on the female partner even using sperm cells from the former husband.
- Cryopreservation: The frozen pre-embryo is the property of the couple alone and may be transferred to the same wife in a successive cycle but only during the validity of the marriage contract.
- Multifetal pregnancy reduction is only allowed if the prospect of carrying the pregnancy to viability is very small. It is also allowed if the life or health of the mother is in jeopardy.
- Surrogate motherhood: The present situation in the Islamic world is that surrogacy is forbidden.

Sunni Islam is the dominant form of Islam throughout the Muslim world. 80%–90% of the world's Muslims are Sunni, and more than 90% of Egypt's citizens are Sunni Muslims. Shi'a is the minority branch of Islam, but it predominates in Iran, parts of Iraq, Lebanon, Bahrain, Syria and Saudi Arabia, as well as Afghanistan, Pakistan and India.

There was no separation between Sunni and Shi'a Islam on these principles until the late 1990s. The Supreme Leader of the Shi'a branch of Islam then issued a proclamation which effectively permitted the use of donor sperm and eggs. This was very significant in opening opportunities for treatment. Lebanon is largely Shi'ite, and many IVF clinics in Lebanon now offer treatment to their own citizens but also to members of other Muslim Middle Eastern countries where the prevailing thinking is Sunni and less permissive of donor services (14).

In summary, a Muslim couple presenting with infertility is likely to experience cultural pressure to have children in addition to their own personal desire to do so. Their acceptable options for assisted conception will be determined by whether they are Sunni or Shi'a Muslims. By understanding the difference in the acceptable treatment options, clinicians can better serve these communities.

10.2.3 Christianity

10.2.3.1 Orthodox

The Eastern Orthodox Church consists of several large independent and self-governing churches, including the Orthodox Churches of Russia, Greece and Romania. While each of these has different positions on some areas, their position on *in vitro* fertilisation, on assisted conception and on abortion are quite consistent and well defined. The Orthodox Church's position is that the status of the embryo is not dependent on its stage of development (15). The fertilisation of an egg is regarded as a significant and defining act, and following it, the characteristic of a new human life has been determined. The embryo, from conception, is regarded as having a soul, and is considered to have rights to its own identity, to life and to eternity.

These rights inevitably cause tensions and difficulty when practicing assisted conception. There are minor differences between the national churches as to what is acceptable in this area, but the overarching Orthodox Church position is (16):

- It does not approve assisted reproduction as the solution to infertility (preferring couples to seek less invasive or intrusive methods).
- It does not approve of spare leftover embryos.

- It does not approve of multiple embryos being transferred if that will lead to selective reduction of some of the fetuses.
- Greek and Russian churches are publicly flexible on the use of IVF when it is used between a man and a woman within a marriage.

An observant Orthodox couple will want to respect the uniqueness of an embryo, but may be willing to consider assisted conception within a marriage relationship. They are likely to prefer simple treatments but clearly deeper enquiry will clarify this.

10.2.3.2 Catholicism

The formal position of the Catholic Church is that external interference in the process of reproduction is not permitted (17). Like the Orthodox Church, the Catholic Church teaches that the embryo is regarded as having a soul from conception.

Catholic teaching on reproduction and reproductive medicine is generally conservative:

- The Church does not approve of IVF and of artificial insemination with the husband's sperm because they separate the unifying and reproductive meaning of sexual intercourse.
- The Church is opposed to extracorporeal conception whereby embryos are formed as a result of a technological action rather than sexual intercourse, *and subject to quality control* (which means they would be discarded if not of sufficient quality) (17).

The use of a sperm-friendly condom during intercourse to collect sperm would generally be acceptable. The use of donor gametes is not permitted because the child conceived using donor sperm or egg is not the child of the marriage union; like surrogacy, the use of donor gametes breaks the marriage union and is not permissible. Access to treatment by single and lesbian women is not permitted, as this approach will be deliberately depriving children of the chance of being raised by their genetic fathers (17).

Clinicians treating Catholic couples for infertility should be sensitive to the teaching of Catholic Church on these matters, but also realise that many Catholic couples form their own views on these issues!

10.2.3.3 Protestantism

There is not a single Protestant stance on reproduction and infertility treatments. Within the broad group of churches, individual positions will be based on Biblical interpretation. Differing religious groups within Protestantism espouse views ranging from quite liberal to more restrictive. On matters like IVF and assisted conception, institutions, such as the Church of England, publish statements of agreement from experts. Generally, the position permits IVF and other forms of direct patient treatment, including donor gametes, but is more cautious when mitochondrial transfer and similar treatments are involved. Surrogacy is also an area which is not approved (18).

A 1995 paper of the status of the early embryo (19) has not been challenged by scientific progress. People with a Protestant faith will hold a wide range of opinions (similar to most other faith groups). Some seek help to rectify whatever is found wrong within their own bodies but, if they still remain unable to conceive, will hold this to be God's will. Some will feel able accept the use of donor sperm and insemination, based on the concept of 'grafting in' outside people to make God's family, but they will refuse anything which creates embryos outside the body. There will be some who are comfortable with the production of embryos *in vivo* provided all the embryos are replaced in the uterus if viable. That 1995 paper accommodates the realities of identical twinning, hydatidiform moles and other disorders in early human development, but serves to maintain the high status of the embryo once implanted. In that paper, it was proposed that the only acceptable fate for 'spare' embryos was to be cryopreserved and used in the future for the couple who were the parents of the embryos, and not to be used as commodities.

When treating devout Protestant couples, clinicians should therefore be aware that many will struggle with using IVF because of embryo wastage will struggle with using assisted conception because of embryo wastage, and some will struggle with donor insemination, sharing much of the beliefs on donor gametes being an intrusion into the relationship with Catholics and Orthodox believers.

10.2.4 Judaism

Judaism is not a single entity, but is divided in three main groups, Orthodox Jews, conservative Jews and reformed Jews (12). The world's estimated 14 million Jews are approximately divided respectively as 10%, 5% and 85%. Amongst Jews, fertility and children are seen as desirable, arising from the instruction by God to 'be fruitful and multiply' (Genesis 1: 28), and reflected in several biblical accounts of women joyfully giving birth after long periods of waiting and being infertile. Judaism considers that 'ensoulment' does not occur until 40 days after conception, so that assisted reproductive techniques using the couple's own gametes are acceptable, as are techniques such as PGD and PGS. There is some disquiet about obtaining semen by masturbation, often amongst Orthodox Jews, because there are injunctions in the Bible not to do this. Using nonspermicidal condoms is generally considered an acceptable way around this. Given the variety of viewpoints within this tradition, it remains advisable to ask questions of the couple, but with the awareness that donor sperm or eggs, and surrogacy are the subject of some debate and may cause problems for the couple in front of you.

10.3 PLANNING CARE WITH COUPLES TO INCORPORATE THEIR VALUES AND CULTURAL BACKGROUND

Once the initial history taking and examinations are completed, and investigative results are available, a plan for care should be agreed. As part of this, it would be wise and sensible to ask a couple:

- Do you have any reservations or concerns you want to raise because of your beliefs?
- Would you like to talk about those concerns now before we move on to allow us to incorporate them in how we care for you?

You may then be alerted that particular treatments might not be acceptable. This might mean that low-dose IVF might be the best option for a couple if they only want enough embryos for a single fresh transfer, with none left over; collection of sperm might require a special nonspermicidal condom to collect it during intercourse. These are *all* reasonable adjustments to make.

10.4 REPRODUCTIVE MEDICINE IN DIFFERENT ENVIRONMENTS

Practicing reproductive medicine in developing country environments is challenging. If you come from a developed country, the familiar clinical supports in outpatient services will be absent, which means that even the simple task of seeing a woman or a couple, taking a history and examination and moving to investigations is more complex. If the doctor is not a native of the region or country, there may be a language barrier to overcome. Investigations may already have been performed which are difficult to interpret, as you may not know when in the menstrual cycle the tests were done, and reference ranges may be alien. Factors for the conversion of common endocrine examinations are included in Table 10.1.

10.5 PERSONAL EXPERIENCE

In sub-Saharan Africa, where I have had limited experience, obesity is widespread among women. This is reported on by several investigators, usually recognising the juxtaposition between findings in the population of both starvation and obesity and of overfecundity and infertility. The clinical impact of this obesity is that many women suffer from disorders of ovulation, with and without the

Table 10.1 Useful conversion factors for results (read in conjunction with Table 3.1).

Hormone	SI units	Multiply by this	Imperial/US/conventional
Oestradiol	pmol/L	0.2724 -> <- 3.6713	pg/mL
Progesterone	nmol/L	0.3145 -> <- 3.1801	ng/mL
Prolactin	mIU/L	0.047 -> <- 21.2766	µg/L
Testosterone	nmol/L	0.2884 -> <- 3.4672	ng/mL
Sex hormone-binding globulin	nmol/L	0.095 -> <- 10.5263	µg/mL

manifestations of polycystic ovary syndrome. The infrequent or absent ovulation that accompanies obesity is treatable by weight loss, but this is not easy when the available food is mostly starch based with little fruit or vegetables), and by simple medical interventions (oral ovulation induction agents, oral weight reduction medication – simple but often not readily available in poorly resourced counties).

A second condition affects in women in sub-Saharan Africa is fibromyomata (uterine fibroids). Fibroids are amenable to medical treatment in the short term, although these medications are expensive, but long-term management requires either open or laparoscopic myomectomy. This surgery has a 1%–2% risk of uncontrollable bleeding, and the possibility of an emergency intraoperative hysterectomy. The surgical skills required for this surgery are limited to a small number of reproductive surgical specialists (20).

A final major negative contributor is the behaviour and attitudes of men to the fertility problems of their wives/partners. In attitudes, infertility is a curse and a blight on a family if a woman cannot have a baby. The challenge for infertile women in the developing world is that their men will rarely attend with them when they see the doctor and trying to get a sperm sample can be quite difficult. This is unfortunate, given that up to 50% of infertility problems are related to problems on the male side (21). In developing countries, large numbers of men smoke tobacco, but in addition will ingest other substances such as khat. Tobacco has well-recognised negative effects on sperm quality and fertilisation, but so does khat.

Khat, *Catha edulis,* grows wild or is cultivated in East Africa and southern Arabia (25). Chewing the leaves and young branches provides rapid stimulation, and addiction causes loss of appetite, stomatitis, gastritis, gastric ulcers and constipation accompanied by haemorrhoids. Chemically, the leaves contain alkaloids, including cathinone, a psychoactive alkaloid chemical that stimulates the central nervous system (22).

Khat has mild hallucinogenic effects, inducing euphoria. It also has detrimental effects on sexual function and desire. These effects appear to be mediated by prolactin secretion, which leads to Leydig cell suppression and reduced endogenous testosterone production, affecting both spermatogenesis and potency. In addition, it also has negative effects on sperm function, though its measurable impact is not as great as tobacco. Animal studies show dramatic effects on sperm analysis measurements, with reduced sperm motility and sperm as well as serum testosterone levels (23).

10.6 CONCLUSION

Assisted conception methods are not the only way to achieve a pregnancy, although moving couples straight from presentation might be attractive, but it is more interventive and it still carries unquantified risks (see Chapter 8, Section 8.7.3) (24). The less exciting methods of investigation, diagnosis and medical and surgical managements have been allowed to fade into the background, because generally this is a slower process and results are slower to appear – these methods do work, however, and are generally less costly and risky than assisted conception techniques.

Infertility seems to continue to be a 'woman's problem' in most settings around the world – and while the evidence presented in this chapter and book will, hopefully, have persuaded the reader that that is far from true, and causes are evenly balanced; the stigma of infertility does still tend to fall on the woman.

World religions tend to be restrictive on what is permitted within the range of treatment options for infertile couples. Generally, for a couple in this situation, discussions with their religious leader (imam, priest, rabbi, other) may assist in finding ways forward through this, and as clinicians, we need to be sensitive to those pressures which we may not understand ourselves.

REFERENCES

1. Ombelet W. Global access to infertility care in developing countries: A case of human rights, equity and social justice. *Facts, Views & Vision in ObGyn*. 2011;3(4):257–66.
2. Inhorn MC, Patrizio P. Infertility around the globe: New thinking on gender, reproductive technologies and global movements in the 21st century. *Human Reproduction Update*. 2015;21(4):411–26.
3. Ombelet W, Cooke I, Dyer S, Serour G, Devroey P. Infertility and the provision of infertility medical services in developing countries. *Human Reproduction Update*. 2008;14(6):605–21.
4. Mascarenhas MN, Flaxman SR, Boerma T, Vanderpoel S, Stevens GA. National, regional, and global trends in infertility prevalence since 1990: A systematic analysis of 277 health surveys. *PLoS Medicine*. 2012;9(12):e1001356.
5. Collins JA, Burrows EA, Wilan AR. The prognosis for live birth among untreated infertile couples. *Fertility and Sterility*. 1995;64(1):22–8.
6. Lunenfeld B, Van Steirteghem A. Infertility in the third millennium: Implications for the individual, family and society: Condensed meeting report from the Bertarelli Foundation's second global conference. *Human Reproduction Update*. 2004;10(4):317–26.
7. Fathalla MF, Sinding SW, Rosenfield A, Fathalla MM. Sexual and reproductive health for all: A call for action. *Lancet (London, England)*. 2006;368(9552):2095–100.
8. Bahamondes L, Makuch MY. Infertility care and the introduction of new reproductive technologies in poor resource settings. *Reproductive Biology and Endocrinology: RB&E*. 2014;12:87.
9. Ombelet W. Is global access to infertility care realistic? The walking egg project. *Reproductive Biomedicine Online*. 2014;28(3):267–72.
10. Coward HG, Lipner J, Young KK. *Hindu Ethics: Purity, Abortion, and Euthanasia*. Albany, NY. SUNY Press, 1989.
11. Bhattacharyya S. *Magical Progeny, Modern Technology: A Hindu Bioethics of Assisted Reproductive Technology*. Albany: SUNY; 2006. 162 p.
12. Sallam HN, Sallam NH. Religious aspects of assisted reproduction. *Facts, Views & Vision in ObGyn*. 2016;8(1):33–48.
13. Serour GI, Aboulghar MA, Mansour RT. Bioethics in medically assisted conception in the Muslim world. *Journal of Assisted Reproduction and Genetics*. 1995;12(9):559–65.
14. Inhorn MC. Making Muslim babies: IVF and gamete donation in Sunni versus Shi'a Islam. *Culture, Medicine and Psychiatry*. 2006;30(4):427–50.
15. Koios N. Embryo and foetus as seen by Orthodox Church. *Periodicum Biologorum*. 2009;111(3):359–63.
16. Nikolaos M. The Greek Orthodox position on the ethics of assisted reproduction. *Reproductive Biomedicine Online*. 2008;17(Suppl 3):25–33.

17. Ford NM. A Catholic ethical approach to human reproductive technology. *Reproductive Biomedicine Online*. 2008;17(Suppl 3):39–48.
18. Sutton A. Three Christian views on assisted conception and marriage – The Roman Catholic Church, Church of England and Presbyterian Church of Scotland. *Eubios Journal of Asian and International Bioethics*. 1996;6:105–7.
19. Cahill DJ. Issues of Early Life. In: Beer D, editor. *Christian Choices in Healthcare*. Leicester: Inter-Varsity Press; 1995.
20. Dedes IA, Squires RM, Thomas EK, Mills MS, Cahill DJ. Should we still be undertaking open myomectomies? A five-year retrospective case review. *Gynaecology and Perinatology*. 2016;1(1):9–17.
21. Dhont N, van de Wijgert J, Coene G, Gasarabwe A, Temmerman M. 'Mama and papa nothing': Living with infertility among an urban population in Kigali, Rwanda. *Human Reproduction (Oxford, England)*. 2011;26(3):623–9.
22. el-Shoura SM, Abdel Aziz M, Ali ME, el-Said MM, Ali KZ, Kemeir MA et al. Deleterious effects of khat addiction on semen parameters and sperm ultrastructure. *Human Reproduction (Oxford, England)*. 1995;10(9):2295–300.
23. Nyachieo A, Kiraithe MM, Spiessens C, Chai DC, Kiulia NM, D'Hooghe TM et al. Short-term effects of high-dose khat on sperm parameters and reproductive hormonal levels in olive baboons (*Papio anubis*). *Gynecologic and Obstetric Investigation*. 2013;75(2):109–14.
24. ESHRE Capri Workshop Group. Economic aspects of infertility care: A challenge for researchers and clinicians. *Human Reproduction (Oxford, England)*. 2015;30(10):2243–8.
25. Krikorian AD. Kat and its use: An historical perspective *Journal of Ethnopharmacology*. 1984;12 (2):115–78.
26. Ombelet W. The development of in-vitro fertilization in Africa. In: *In Vitro Fertilization: The Pioneers' History*. Eds: Kovacs G, Brinsden P, DeCherney A. Cambridge, Cambridge University Press, UK, 2018, p256.
27. Hammarberg K, Kirkman M. Infertility in resource-constrained settings: Moving towards amelioration. *Reproductive Biomedicine* (Online). 2013;26 (2):189–95.

Reproductive biology in one other great ape (the gorilla) 11

OVERVIEW

This chapter discusses the challenges of making reproductive medicine diagnoses in other large mammals and the interpretation of any investigations and treatments that might follow. Some comments are included on the progress and outcomes of pregnancy.

11.1 GRAPPLING WITH PHYSIOLOGY

Getting involved in the medical care of great apes is an honour, a privilege and a responsibility. Gorillas are a critically endangered species. There are still about 100,000 Western lowland gorillas in the wild and maybe 4000 in zoos. Medical invitations to take part in their care are unusual and daunting. How does a career in great ape gynaecology start? Like many other careers, by accident.

This story begins in 2004, with Salome, a 29-year-old Western lowland gorilla, who had been mating regularly with the principal male gorilla, Jock, in Bristol Zoo, but was not getting pregnant. Salome had previously had a baby conceived naturally, born in 1988, when she was 16. Jock had fathered several other offspring. The veterinary surgeons in the zoo decided to ask for help with her infertility. Bristol Zoo had links with the University of Bristol Veterinary School which eventually led them to ask me to help.

Following initial discussion and history taking (as far as possible), any recent blood tests were reviewed, and these showed nothing untoward and very little useful except that the gorilla wasn't anemic. Sample collections for hormone assays were planned, with the option of considering some more invasive tests requiring anaesthesia such as a transvaginal scan and a hysterosalpingogram.

Sampling for hormones in great apes is not a simple task. Some gorillas have been trained to put out their finger for finger pricking (in return for a treat), but Salome had never managed to do that. As a result, the zoo staff had to obtain urine samples. Even that was complex. Each gorilla will have their own space in the compound to which they retreat if unwell, as well as for feeding and sleeping. Each such space has a sloped floor leading to a central hole to collect all the effluent of the day, generally urine. This is how urine samples were collected from Salome and sent off for analysis (in Hamburg, Germany). The results came back expressed as urinary concentrations of oestrone conjugates (E1C) and pregnanediol-3-glucuronide (PdG). In humans, at least, these markers are considered accurate assessors of ovulatory function (1).

In particular, rising progesterone levels always follow ovulation, and indicate that it has occurred. Single measures of E1C and PdG levels could not tell us much, so urine samples were collected regularly and sent for analysis, allowing the results to be plotted against Salome's cyclical oestrus cycle, as recorded by the zookeepers. Menstrual cycle activity in gorillas differs slightly from other great apes. They do not menstruate in a discernible manner, though blood can be detected in urine when it (menstruation) occurs, and this lasts 2–3 days. Oestrus is indicated by the desire to mate and is associated with a rise in oestrogen levels generally, indicating the temporal proximity of ovulation. In gorillas, oestrus is accompanied by minor swelling of the labia at the introitus, something the zookeepers are able to note and record. Labial swelling usually disappears when the LH surge happens (and when ovulation occurs) (2).

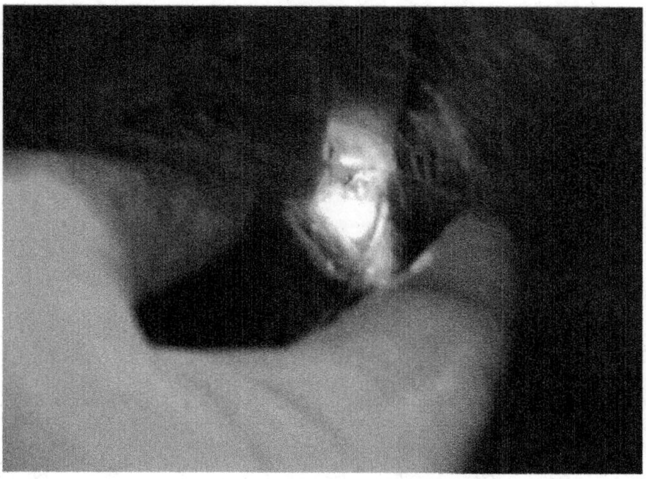

Figure 11.1 Examination of the vaginal orifice prior to scan and x-ray (the diameter is no bigger than my thumb). (By courtesy of Bristol Zoological Society.)

In Salome's case, over 7–8 months of investigation, it was evident that there were no signs of any alteration in progesterone hormone metabolites in the urine. This led to the conclusion that Salome was not ovulating - and thence to consider what means of ovulation induction could be used. There were two choices - oral (mild) stimulation using clomifene citrate (discussed extensively in Chapter 5, Section 5.3) and injectable gonadotrophins containing FSH from urinary or recombinant sources (also in Section 5.3).

While urinary hormone investigations were proceeding, more invasive tests were undertaken to examine her pelvis by ultrasound transvaginally and to undertake a hysterosalpingogram. From the contemporaneous notes

> Following induction of anaesthesia (with 4 mg Medetomidine and 600 mg Ketamine by dart and after the plain X rays of the gorilla's pelvis had been taken, I undertook a transvaginal scan (the diameter of the vagina appeared unexpectedly small - Figure 11.1). On transvaginal scanning the uterus was visualised. The uterus looked normal with no evidence of intrauterine or myometrial abnormalities. The endometrium was normal in thickness and measured 7–8 mm. Longitudinal and transverse views were taken of the endometrial cavity.
>
> We then proceeded to a hysterosalpingogram. A size 8 once-only use catheter was inserted in the cervical cervix and access to the cervical canal was achieved with particular difficulty (the cervix was very flush with the vagina). The vaginal diameter was approximately 2 cms and the cervix was normally placed at the distal end of the vagina, about 10 cms from the introitus. Following the identification of the canal, the catheter was passed through it to a depth of 6–8 cms. A number of views of the uterine cavity were taken. The uterine cavity was normal in outline and on several views, the left fallopian tube and distal end of the fallopian tube were clearly outlined. On no picture was the right fallopian tube identified and therefore right tubal patency was not confirmed.
>
> *Assessment*
>
> The uterine cavity has been determined to be normal both ultrasonographically and radiologically (any description of normal means that the findings were normal compared to a human female). The left fallopian tube is entirely normal. The right fallopian tube was not identified. This finding is more likely to be attributed to spasm at the cornual end of the tube rather than to a true tubal blockage. It is very clear from this assessment that there is no need for surgical intervention. Fertility is only reduced by 30% if one tube is blocked, and it is more likely that this finding is co-incidental and related to spasm than to a true fallopian tube cornual blockage.

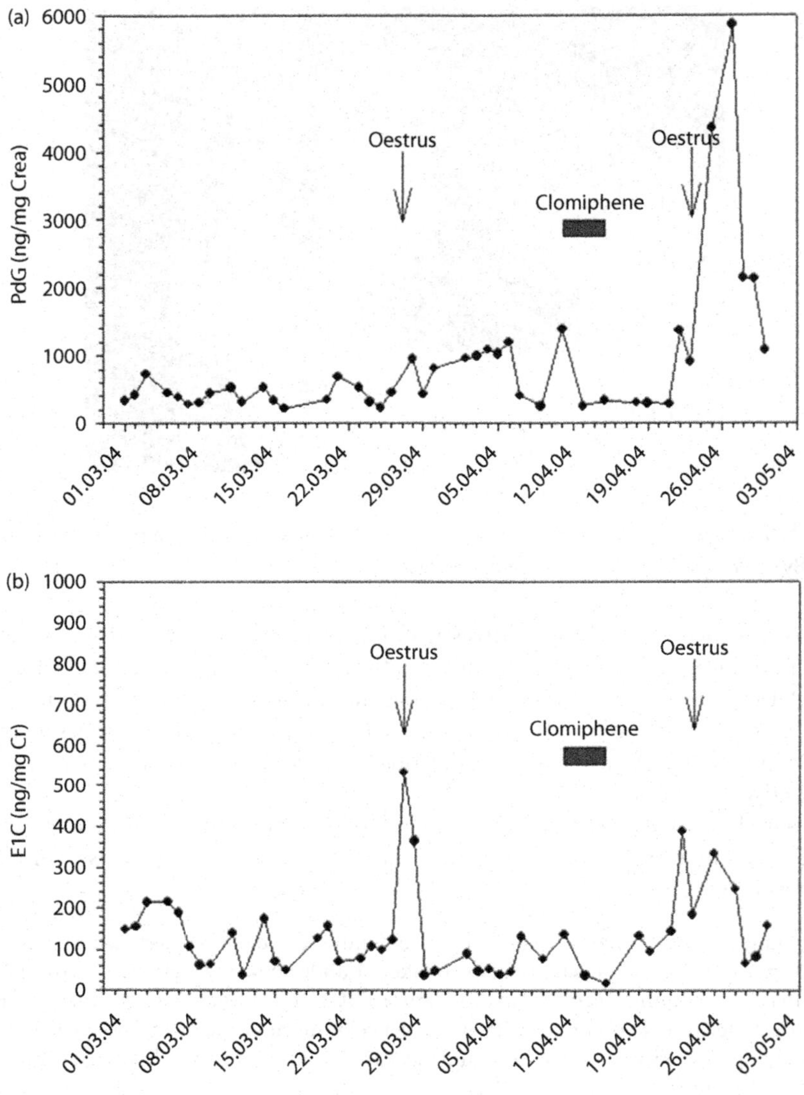

Figure 11.2 Conjugates of (a) progesterone and (b) oestradiol measured in daily samples of urine. (By courtesy of Bristol Zoological Society.)

These assessments were undertaken in the gorilla enclosure, on the floor, and with the alpha male Jock on the other side of a (admittedly steel) door, making it clear vocally that he did not appreciate this 'assault' on a member of his harem. It was not a pleasant experience!

The simpler approach to ovulation induction was chosen, recognizing it was less effective, and she was started on clomifene citrate (Clomid; Sanofi, Guildford, UK). Because Salome's weight exceeded 150 kg, she was given 100 mg daily (twice the normal human starting dose). The effect of this administration is seen in Figure 11.2 – the upper slide shows the data for pregnanediol-3-glucuronide. Here oestrus is noted twice; progesterone levels do not rise after the first oestrus, clomifene is administered and shortly after the second oestrus, a major spike in progesterone levels was noted, indicating ovulation. Clearly gorilla physiology is not identical to human physiology, as the progesterone surge occurred immediately after oestrus whereas in humans, recall that the progesterone surge occurs a whole week after ovulation. Gorillas are similar to humans in many ways – but it is foolish to assume they are identical. Clomifene was then continued for several months, and she conceived.

Unfortunately, there was an unexpected fetal death at 5 months, and Salome delivered a stillborn baby. After a period, the medication was commenced, and she again conceived in early 2006.

Gestational length in gorillas is not precise, but a variety of estimates put it between 256 and 270 days (2) (not dissimilar to the human's 280 days). The 2006 pregnancy of Salome was expected to finish in December, probably just before Christmas Day. This pregnancy was straightforward until the very end when Salome showed signs of being unwell and was proteinuric. Given that she had lost one baby and given the deterioration in her physical health, the vets and her carers requested that I attend on the morning of 19 December 2006, to undertake a caesarean section. Thankfully, I was told on that morning that she had delivered during the night and that all was well. I say "thankfully" because mother and her male baby, Komale, were fine and did recover, but it was interesting to note that Salome's kidneys shut down and she was anuric from Friday to the following Tuesday – 5 days in total. I was very grateful that we did not deliver her by caesarean section, as that insult and the anuria might have led to severe morbidity for her, maybe even mortality.

Interestingly, she conceived spontaneously 5 years later (September 2011) at the age of 35 and had a completely straightforward pregnancy and delivery of a girl.

Her mother, Salome, had conceived spontaneously. Her father was Komale, whose conception I had contributed to medically 9 years earlier. The pregnancy was uncomplicated until Salome's 38th week (approximately). Then, she became withdrawn, ate very little and stayed in her enclosure. Proteinuria was noted when her urine was tested. Given this change, I attended and scanned her uterus and found the fetus to have almost no amniotic fluid, to have a heart rate of 100 beats/minute and to be unresponsive to stimuli. All these pointed towards a fetus in difficulty (if it were human) and so we proceeded to deliver her by caesarean section that day. Once the anaesthetic wore off, the baby thrived, though it needed a foster mother (one of the other adult females). Salome's diagnosis was unresolved: the initial thoughts had been pre-eclampsia. The proteinuria continued for 6 months with haematuria, which eventually responded to corticosteroids, which was more suggestive of some glomerulonephritis.

All the excitement and media attention was focussed on the delivery of Afia in 2016 (Figure 11.3). My second contact with gorilla biology occurred in February 2016. A female gorilla, Kera, conceived spontaneously. The pregnancy was uncomplicated until the (presumed) 38th week. Then she became withdrawn, ate very little and stayed in her enclosure all the time. Urine testing revealed proteinuria. I was asked to attend and scan her uterus. She was anesthetized in her enclosure using ketamine and transferred to the (small) zoo theatre. There, she was intubated and connected to monitoring equipment. Her blood pressure was 'normal'. Abdominal scanning showed that the fetus had almost no amniotic fluid, the heart rate was 100 beats/minute, which in a human baby is worryingly low; and the fetus was unresponsive to stimuli (prodding). All the evidence pointed towards a fetus in difficulty (again, utilising human criteria). The working diagnosis was pre-eclampsia, despite normal blood pressure, and we proceeded to deliver her by caesarean section. The abdomen was shaved and prepped, and efforts were made to catheterise her. After much fumbling, the urethral opening was located internally, on the anterior wall of the vagina (2).

The approach used was a Pfannenstiel incision, anticipating that the uterus would have a lower segment. Only afterwards, I learned that the pregnant gorilla uterus does not always have a lower segment! The skin was extremely tough to cut, and separating the rectus muscles required considerable force. The lower segment was identified and opened. The baby's head was delivered by forceps which the zoo made available.

The female baby was unresponsive and floppy, but alive. Apgar scores at 1 and 5 minutes were 0. She was oxygenated and received chest compressions for the next hour or so, until she woke up, at approximately the same time as her mother came out of the anaesthetic! At 2.4 kg, she had experienced growth retardation; the usual gorilla birth weight at term ranges from 3–4 kg. Once awake, she fed normally and enthusiastically from a bottle.

Returning to Kera, the uterus and abdominal muscle were closed as usual, and subcuticular stitches were used for the skin. At the request of zoo staff, I placed 'distraction stitches' in her

Figure 11.3 Holding human and animal offspring – granddaughter and Afia. (By courtesy of Bristol Zoological Society and Elizabeth Ewing.)

left forearm, stitches with long threads hanging out – to occupy the gorilla while the abdominal wound healed. Assuming Kera was suffering from preeclampsia, I expected her to recover quickly after delivery. She didn't. She remained lethargic and off her food, proteinuria continued and some samples showed haematuria. Veterinary medicine appears to use corticosteroids on an empirical basis, and Kera was put on a 6-month course, by the end of which she was fully recovered and back to her healthy self. The unproven cause for her illness was that she had glomerulonephritis.

Finally, many of the important facts have been published already (3), but the key things to be aware of in operating on gorillas and large apes (which differ from humans) in late pregnancy are:

- The urethra being interior and on the anterior aspect of the vagina, 2–3 cm or so upwards, not external like it is in humans.
- The flatter saucer shape of the gorilla pelvis (as opposed to human conical) means that the lower segment is often poorly formed or nonexistent. That absence meant that for others undertaking caesarean sections, the uterus was incised through the upper segment, as the lower segment had not developed.
- Second caesareans are likely to be more difficult that the first, at least in the experience of the only person that I know who of that has done this (Dr Penny Foster, Melbourne), who found that the abdominal approach at second caesarean was greatly complicated by the presence of adhesions (4).
- Bring equipment you are familiar with – most zoos do not routinely do laparotomies in animals over 100 kg, so you may struggle to find equipment you are familiar with.
- Encourage the use of a different agent to ketamine in the case of a caesarean section, as no specific antidote to it presently exists, and in the case of the caesarean section I undertook, the offspring had prolonged respiratory suppression, probably because of the ketamine.

11.2 SUMMARY

This was one of the most exciting times of my career and with lots of unexpected media exposure. It was great fun, but the day-to-day care and long-term commitment of the vets and the gorilla enclosure team does not get noticed in the media hype. I am very grateful to have had to witness it.

REFERENCES

1. Venners SA, Liu X, Perry MJ, Korrick SA, Li Z, Yang F et al. Urinary estrogen and progesterone metabolite concentrations in menstrual cycles of fertile women with non-conception, early pregnancy loss or clinical pregnancy. *Human Reproduction (Oxford, England).* 2006;21(9):2272–80.
2. Graham CE. *Reproductive Biology of the Great Apes Comparative and Biomedical Perspectives.* Academic Press; 1981. 437 p.
3. Walter S, Ali A, Killick R, Day C, Barrows M, Cahill DJ. Caesarean delivery of a Western lowland gorilla with fetal distress and suspected pre-eclampsia. *International Journal of Reproduction, Contraception, Obstetrics and Gynecology.* 2018;7(7):2933–6.
4. McCracken H, McBain JC, Foster PA, Moriarty K, Campbell NT, editors. Two successive cesarean deliveries in an elderly primigravid lowland gorilla (*Gorilla gorilla*) and neonatal care of the offspring. *Proceedings American Association of Zoo Veterinarians.* 1994. Pittsburgh, PA.

Research questions still unanswered – And further reading

12

QUESTIONS ARISING FROM THE TEXT OF THIS BOOK THAT HAVE NOT BEEN FULLY ADDRESSED (NOT EXCLUSIVE)

Uterine structural anomalies – Do women with congenital anomalies and controls have different outcomes in natural or assisted reproduction cycles?

Uterine structural anomalies – Do women with congenital anomalies benefit from surgical intervention in terms of clinical pregnancy or live birth rates?

Fibroids – Outcomes and complications of uterine artery embolisation.

Fibroids – Which fibroids have an impact on fertility – and by how much?

Fibroids – Removal by open surgery or laparoscopically?

Fibroids – Outcomes and complications of surgery (open or laparoscopic) to include caesarean section at next delivery, if successful in conceiving.

Fibroids – is there potential for developing a vaccine to prevent fibroid growth by attenuating some strain of HPV?

Adenomyosis – Effect on fertility, if any? If yes, optimal management.

Adenomyosis – Optimal diagnostic methods?

PCOS – Developing a model for its effects which addresses all or most of the loose ends.

Endometriosis – Does endometriosis diminish the likelihood of getting pregnant (at fertilisation, pregnancy or population level)?

FURTHER READING

Balen A. *Infertility in Practice.* 4th edn. CRC Press, Boca Raton, FL; 2015.

Epstein R. *Human Molecular Biology.* 1st edn. Cambridge, Cambridge, UK; 2012.

Johnson M. *Essential Reproduction.* 8th edn. Blackwell, Oxford, UK; 2018.

McVeigh E, Guillebaud J, Homburg R. *Oxford Handbook of Reproductive Medicine and Family Planning.* 2nd edn. Oxford (particularly for family planning and contraception); Oxford, UK; 2013.

Speroff L, Fritz MA. *Clinical Gynecologic Endocrinology and Infertility.* 8th edn. Lippincott Williams & Wilkins; Philadelphia, Pennsylvania 2011.

Index